Green Energy and Technology

Climate change, environmental impact and the limited natural resources urge scientific research and novel technical solutions. The monograph series Green Energy and Technology serves as a publishing platform for scientific and technological approaches to "green"—i.e. environmentally friendly and sustainable—technologies. While a focus lies on energy and power supply, it also covers "green" solutions in industrial engineering and engineering design. Green Energy and Technology addresses researchers, advanced students, technical consultants as well as decision makers in industries and politics. Hence, the level of presentation spans from instructional to highly technical.

Indexed in Scopus.

More information about this series at http://www.springer.com/series/8059

Ali M. Eltamaly · Almoataz Y. Abdelaziz ·
Ahmed G. Abo-Khalil

Editors

Control and Operation of Grid-Connected Wind Energy Systems

Editors
Ali M. Eltamaly 🄳
Sustainable Energy Technology Center
King Saud University
Riyadh, Saudi Arabia

Almoataz Y. Abdelaziz 🄳
Electrical Power and Machines Department
Ain Shams University
Cairo, Egypt

Ahmed G. Abo-Khalil 🄳
Electrical Engineering Department
Majmaah University
Almajmaah, Saudi Arabia

ISSN 1865-3529 ISSN 1865-3537 (electronic)
Green Energy and Technology
ISBN 978-3-030-64335-5 ISBN 978-3-030-64336-2 (eBook)
https://doi.org/10.1007/978-3-030-64336-2

This Springer imprint is published by the registered company Springer Nature Switzerland AG
The registered company address is: Gewerbestrasse 11, 6330 Cham, Switzerland

Preface

Renewable energy systems are becoming more attractive for generating electric power due to their sustainability, environmental friendliness, can be used as a replacement for the dependency on fossil fuels, and they can be localized near the loads to add support to the electric power system. Wind energy is one of the fastest growing renewable energy systems and with the increasing improvements to the control system of wind energy systems, it can increase its penetration to the electric power system mix. The generated energy and the stability of the wind energy systems depend mainly on the wind speed variation and the improvement in the control system of this energy source will help to increase their use for generating electric energy. This book is introduced to present the control and operations of the integration of wind energy systems with electric power systems. The book introduced the size of the excitation capacitor required when using a three-phase induction generator with an accurate model for the losses in the rotor and stator of the self-excited induction generator. Also, the use of a Double-fed Induction Generator (DFIG) as a generator in wind energy applications is getting more attention in many chapters of this book. The control of generated power in the DFIG and maximum power tracking is introduced and discussed in many chapters of this book. The power quality and harmonics injected into the power system are introduced and discussed. Many techniques to overcome (treat) the effects of harmonics on the power system due to the integration of renewable energy systems to the electric utility are introduced and discussed. The communication techniques used in the control and operation of the wind energy system and the SCADA systems are introduced to show how they can control the wind energy systems. Many chapters have been introduced in the book to show the effect on the power system when the wind energy system is used as a distributed generation source. Also, smart optimization techniques have been used in one chapter for performance and control improvement and the maximum power point tracking of the wind energy system. Moreover, planning, design, and cost analysis of the wind energy system and economic analysis of the wind energy system are introduced in many chapters and how they can be used in a hybrid system with other renewable energy sources is introduced and discussed. The selection of wind turbine suitable for sites and the criteria used to choose the best sites are introduced in detail.

This book is introducing a good overview on how to use the wind energy system working standalone and integrated with the electric utility. These subjects introduced in this book will help students, researchers, designers, and decision-makers for a better understanding of the operation and control of wind energy systems and how it can be interconnected with an electric utility and the financial details of installing wind energy systems. Moreover, this book will be very interesting for the readers who are looking for using wind energy systems to feed loads in isolated areas as well as on the utility scale. It will also help them to know the wind energy systems' characteristics, modeling, operation, challenges, maximum power tracking, and practical implementation.

Riyadh, Saudi Arabia   Ali M. Eltamaly
Cairo, Egypt   Almoataz Y. Abdelaziz
Almajmaah, Saudi Arabia   Ahmed G. Abo-Khalil

Acknowledgment

The editors of this book would like to thank the authors and reviewers for their contributions and efforts. Moreover, we would like to thank all colleagues from K. A. CARE Energy Research and Innovation Center, Riyadh, Saudi Arabia for their help and efforts.

Contents

About the Editors

Ali M. Eltamaly (Ph.D.–2000) is a Full Professor at Mansoura University, Egypt, and King Saud University, Saudi Arabia. He received B.Sc. and M.Sc. in Electrical Engineering from Al-Minia University, Egypt in 1992 and 1996, respectively. He received Ph.D. in Electrical Engineering from Texas A&M University in 2000. His current research interests include renewable energy, smart grid, power electronics, motor drives, power quality, artificial intelligence, evolutionary and heuristic optimization techniques, and distributed generation. He has published 20 books and book chapters and he has authored or coauthored more than 200 refereed journals and conference papers. He has published several patents in the USA patent office. He has supervised several M.S. and Ph.D. theses worked on several National/International technical projects. He got Distinguished Professor award for Scientific Excellence, Egyptian supreme council of Universities, Egypt, June 2017, and he has awarded many prizes in different universities in Egypt and Saudi Arabia. He is participating as an editor and associate editors in many international journals and chaired many international conferences' sessions. He is Chair Professor of Saudi Electricity Company Chair in power system reliability and security, King Saud University, Riyadh, Saudi Arabia.

Almoataz Y. Abdelaziz (SMIEEE' 2015) He received B.Sc. and M.Sc. in Electrical Engineering from Ain Shams University, Cairo, Egypt, in 1985 and 1990, respectively, and Ph.D. in Electrical Engineering according to the channel system between Ain Shams University, Egypt, and Brunel University, U.K., in 1996.

He is currently a Professor of Electrical Power Engineering at Ain Shams University. He has authored or coauthored more than 350 refereed journals and conference papers and 20 book chapters. He has supervised more than 70 M. Sc. and 15 Ph.D. theses in his research areas, which include the applications of artificial intelligence, evolutionary and heuristic optimization techniques to power systems power system operation, planning, and control.

Dr. Abdelaziz is the Chairman of IEEE Education Society chapter in Egypt, a Senior Editor of Ain Shams Engineering Journal, Editor of Electric Power Components & Systems Journal, Editorial Board member, Editor, Associate Editor, and Editorial Advisory Board member for many international journals.

He is also a senior member in IEEE, a member in IET, and the Egyptian Sub-Committees of IEC and CIGRE'. He has been awarded many prizes for distinct researches and for international publishing from Ain Shams University, Egypt.

Ahmed G. Abo-Khalil received the Bachelor and Master of Science in Engineering from Assiut University, Egypt, and Ph.D. from the School of Electrical and Computer Engineering, Yeungnam University, South Korea, in 2007. In 2008, he joined Rensselaer Polytechnic Institute, NY, USA, as a postdoc researcher and worked on a renewable energy project. From 2009 to 2010, he was a Postdoctoral Research Fellow in the Korean Institute of Energy Research, Daejeon, South Korea, working on photovoltaic power conversion systems. In 2010, he moved to Assiut University, Egypt, as an Assistant Professor. He works now as an Associate Professor with the Department of Electrical Engineering, Majmaah University, Almajmaah, Kingdom of Saudi Arabia.

Modeling and Effect of Core Loss in AC Three-Phase Self-excited Generators Used in Wind Energy Applications

Saleh Al-Senaidi, Abdulrahman Alolah, and Majeed Alkanhal

Abstract Green renewable energy sources have been introduced as alternatives to avoid the environmental impact of the hazardous waste of conventional power generation. One of these remarkable exploited green sources is wind energy. Self-Excited AC Generators, namely, three-phase Self-Excited Induction (SEIG) and Reluctance (SERG), generators are used to convert wind power to electric power. Extensive research studies have been carried out on the analyses of dynamic, transient, as well as steady-state performance of these generators. In most of these studies, core losses were neglected. However, different methods have been attempted to consider the core loss by adding resistance to the model of the generators. The values of this resistance are taken as either; (i) fixed, (ii) linearly proportional to the magnetizing reactances (X_m or X_d), or (iii) variable as a polynomial function of X_m (or X_d). This chapter presents a comparative study to assess the above-mentioned three methods under different operating conditions of (SEIG) and (SERG) generators. The effect of neglecting the core loss on the analysis are studied. The method no. (iii) above is used as a reference to evaluate the errors in the other methods. The generator performance resulted from the three methods is shown and compared. Experimental verifications are included to illustrate the most accurate method to account for core loss in generator analysis.

Keywords Core loss · Self-excited · Reluctance generator · Induction generator · Three phase · Steady state · Wind energy

S. Al-Senaidi (✉) · A. Alolah · M. Alkanhal
Department of Electrical Engineering, College of Engineering, King Saud University, Riyadh, Saudi Arabia
e-mail: salih@ksu.edu.sa

A. Alolah
e-mail: alolah@ksu.edu.sa

M. Alkanhal
e-mail: majeed@ksu.edu.sa

© The Author(s), under exclusive license to Springer Nature Switzerland AG 2021
A. M. Eltamaly et al. (eds.), *Control and Operation of Grid-Connected Wind Energy Systems*, Green Energy and Technology,
https://doi.org/10.1007/978-3-030-64336-2_1

1

List of Symbols

a	P.u. speed.
C, X_c	Value of excitation capacitance (μF) and its p.u. reactance at base frequency, respectively.
$C_{\min}$	Minimum excitation capacitance (μF).
V_o	P.u. terminal voltage.
E_q, E_d	P.u. quadrature and direct magnetizing voltages, respectively.
F, u	P.u. frequency and speed, respectively.
I_q, I_d	P.u. quadrature and direct magnetizing currents, respectively.
I_s, I_c, I_L, I_e	P.u. stator, excitation capacitance, load, and core loss currents, respectively.
R_s, R_r, R_e, R_L	P.u. stator, rotor, core loss, and load resistances, respectively.
E_R, E_g	P.u. air gap voltages of SERG and SEIG, respectively.
X_d	P.u. direct axis saturated magnetizing reactance.
X_o, X_m	P.u. unsaturated and saturated magnetizing reactances at base frequency, respectively.
X_q	P.u. quadrature axis magnetizing reactance at base frequency.
X_L, X_r, X_s	P.u. load, and rotor and stator leakage reactances at base frequency, respectively

1 Introduction

The induction and synchronous reluctance machines are considered self-excited when the appropriate value of capacitors are attached to their terminals while the rotors are driven at suitable speeds. The analyses of the dynamic, transient, and steady state of self-excited AC generators have been covered deeply in the literature albeit the core losses have been largely ignored in the studies. In this chapter, the core losses are modeled in a more accurate function, which is associated with the degree of saturation. This association is obtained experimentally and integrated as a nonlinear model of the SERG and SEIG. The performance of the system is obtained by using an optimization technique to solve the nonlinear models. The result is a novel group of curves that accurately describe the performance of the generator when core losses are taken into consideration. Experimental and computed curves match so much that show the accuracy of the presented models.

In SEIG, some publications completely left out the core loss effects [1–10] while others simply used the motor model of the core loss or assumed a constant resistance in shunt with the magnetizing reactance as a replacement for the core loss [11–17]. These approaches are adequate for motor operation because motors operate near the unsaturated region. On the other hand, SEIG operates in the saturated region [1–5] which is largely affected by variations in load, shaft speed, the value of excitation capacitor, and load's power factor. The changes in the level of saturation change the core loss and hence the operating parameters of the generator.

The main advantages of SERG over SEIG are constant frequency operation as well as lower copper and core losses [18–25]. An equivalent circuit similar to a SEIG is presented in [24] to simplify the analysis of a SERG; however, the effect of an essential parameter of the reluctance machine, which is the saliency ratio, was ignored in the analysis [26]. A model was developed for SERG in [23, 25] by applying Park's d-q transformation with consideration of the effect of magnetic saturation and saliency on the performance analysis. Further development of this analysis with variable speed is made in [27, 28] to predict capacitance requirements and operating limits. However, the above studies do not include core losses of the generator in the analyses. An interesting method for analyzing the SERG has been presented in [29] by considering both core losses and saliency. This method is not appropriate for situations where capacitors are connected to the terminals of the generator for self-excitation as the excitation is generated by connecting a rectifier bridge and a DC supply. The analysis in [30] was performed by modifying that in [24], and the core loss and saliency were considered. However, in this analysis, the core loss was assumed to be constant and represented by a constant resistance irrespective of the saturation effects; furthermore, the analysis was valid only for a fixed speed.

On the other hand, good analysis in [31] was developed for the SERG and was verified experimentally by considering the core loss as a variable resistance. The value of the resistance is considered to be directly proportional to the magnetizing reactance, i.e., $R_e = k X_d$, where k is a proportionality constant.

As a matter of fact, any variation or change in the operation conditions, such as the load power factor, load impedance (R_L and/or X_L), speed, and/or excitation capacitance, will definitely vary the level of saturation of the SERG. The saturation-level variation will change the saturated magnetizing reactance, X_d, core loss, and the air gap voltage, and hence the performance of the SERG. For this reason, the core loss needs to be linked to the level of saturation in the SERG by modeling it as a variable parameter.

The aim of this chapter is to propose an accurate model of the core loss that takes into consideration any change in the level of saturation due to changes in speed (u), excitation capacitance (C), load (Z_L), and its power factor (pf) as mention earlier. To achieve this, the core loss resistance is modeled as a variable that is influenced by the level of saturation in the generator. The benefit of this model is that it can be included in the future design of SEIG and SERG to take into consideration the high and varying saturation levels in the generators. In the following sections, the core loss is linked to the level of saturation by modeling it mathematically as a variable function of saturation level that increases the accuracy of the analysis of SEIG and SERG. The accuracy of the proposed core loss model is confirmed by a set of experimental tests that agree with the simulated results.

2   Three-Phase Self-excited Induction Generator

The SEIG is mostly used in isolated power supply and the mode of excitation is achieved by connecting some predetermined optimal value of capacitor bank across the stator terminals. The capacitor banks supply the required reactive power by the induction generator and accordingly corrects the power factor of the machine. Using only the permanent magnetism of the machine, the capacitor bank resonates with the inductance of the machine to build up the voltage of the IG terminals from near to zero to the desired level. Figure 1 illustrates the scheme of the SEIG.

2.1   Analysis

The equivalent circuit in a per phase of a three-phase SEIG with *R-L* load is shown in Fig. 2. Saturation effects are taken into consideration by modeling the core loss resistance R_c, and the magnetizing reactance X_m as variable parameters. The basic parameters of the generator are obtained from the usual DC, locked rotor, and no-load running at synchronous speed tests. From the DC and the locked rotor tests, the values of R_s, R_r, X_s, and X_r are determined. The magnetization curve of the generator was obtained using a no-load running test while the rotor was driven at synchronous speed. This curve reveals the variation of R_c and X_m against the air-gap voltage (or magnetizing current) as shown in Fig. 3. This figure shows that X_m and R_c are varying in relation to the level of saturation in accordance with the air-gap voltage. Figure 4a

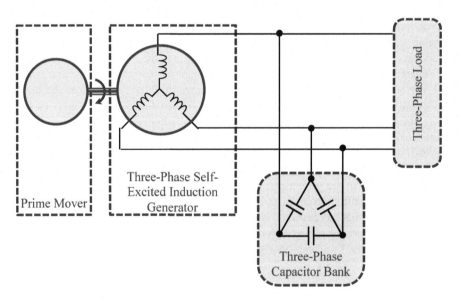

Fig. 1 Three-phase self-excited induction generator

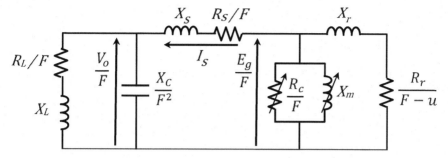

Fig. 2 SEIG per-phase equivalent circuit under the proposed core loss model

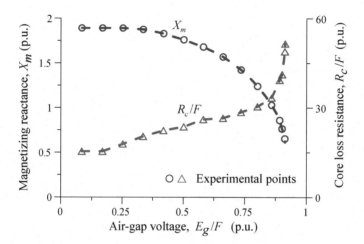

Fig. 3 Variation in R_c/F and X_m against E_g/F in the generator under study

shows a redevelopment of the magnetizing curve shown in Fig. 3 that is essential to obtain the SEIG performance. X_m and R_c are varying because they are considered as variable parameters since they are linked to the saturation level in the generator which varies according to the operating conditions. This fact, to the best knowledge of the authors, is ignored in all the literature of SEIGs [1–9].

2.1.1 Core Loss Modeling

The research gap in the literature as mentioned above is filled by modeling the rate change of the core loss, R_c, with the magnetizing reactance, X_m, as illustrated in Fig. 4b. Experimental results of Fig. 4b shows that the core loss, R_c, varies largely with X_m and can be described mathematically with a fourth-degree polynomial fitted curve. From this model, any change in u, Z_L, and C that leads to the change in the level of saturation can be captured in the change of X_m and consequently in the value

Fig. 4 Variation in E_g/F and $R_c/(F X_m)$ against X_m: **a** E_g/F, **b** $R_c/(F X_m)$

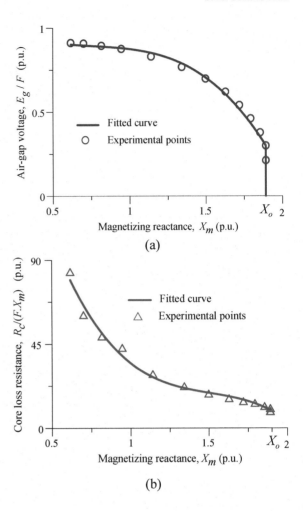

(a)

(b)

of R_c. The curve of the air-gap voltage (E_g) against X_m in Fig. 4a is expressed for computational purposes by a set of piecewise linear functions [1, 32] or by making use of curve fitting with an appropriate degree of a polynomial function as developed by the authors earlier [3].

With the same concept, the variation of the core loss against X_m is also fitted in the same manner as another polynomial function, as shown in Fig. 4b. The fitted curves are expressed as

$$E_g/F = \sum_{i=0}^{n} k_i X_m^i \qquad (1)$$

$$R_c/(F \cdot X_m) = \sum_{i=0}^{r} m_i X_m^i \qquad (2)$$

where m_i and k_i are the coefficients of the polynomials.

The polynomial coefficients are determined experimentally and are given in the Appendix. This method does not alter the characterization given in [1], because the variation in R_c is linked to the variation in X_m (R_c is a function of X_m (i.e., $R_c = f(X_m)$).

2.1.2 Loop-Impedance Solution

In the circuit of Fig. 2, the total impedance, Z_t, across R_c and X_m branch is expressed as

$$Z_t = ((Z_s + (Z_L//Z_C))//Z_r) + Z_m \tag{3}$$

where $Z_s = R_s/F + jX_s$, $Z_L = R_L/F + jX_L$, $Z_r = R_r/(F - u) + jX_r$, $Z_m = (R_c/F)//(jX_m)$, and $Z_c = -jX_c/F^2$.

When the system in Fig. 2 operates at the steady-state condition, the voltage across Z_t is equal to zero [5], i.e.,

$$I_s Z_t = 0 \tag{4}$$

In steady state, there is a passing current in the stator windings (i.e., $I_s \neq 0$), which implies that $Z_t = 0$. From this, the two equations can be expressed as

$$Re(Z_t) = 0 \tag{5}$$

$$Im(Z_t) = 0 \tag{6}$$

Equations (5) and (6) can be solved for two unknowns such as (X_m and F), (X_c and F), (u and F), or (Z_L and F).

Solving the nonlinear equations of (5) and (6) can be done using many schemes that are reported in the literature. A scheme that was presented in [11] used the Newton–Raphson method to solve for the unknowns. Another scheme was given in [1, 2] by rearranging the two equations as two polynomials of high degree in the desired unknown along with the frequency, F. Nevertheless, these schemes cannot fit the proposed model given in this section because the core loss is assumed to be varying. Instead, solving Eqs. (5) and (6) under a variable core loss can be obtained by applying optimization-based schemes, such as the one presented in [3] as described below.

2.1.3   Method of Solution

An optimization-based program was built to solve Eqs. (5) and (6) for the desired unknowns. The desired unknowns can be F and X_m, X_c, Z_L, or u that are obtained by minimizing the value of Z_t (i.e., $|Z_t| = 0$). The system performance of Fig. 2 can be obtained once the unknowns are found with the help of the magnetization curve.

The flowchart of the optimization-based program that solves for X_m and F against the speed is shown in Fig. 5. For obtaining other unknowns such as (F and u), (F and

Fig. 5 Optimization program flowchart to obtain the performance of the SEIG against the speed ($\varepsilon = 1 \times 10^{-5}$)

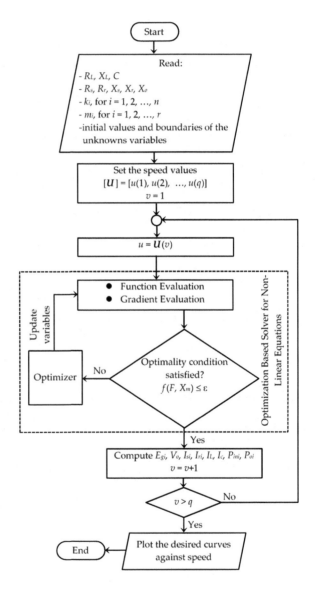

Z_L), and (F and X_c), similar programs were developed. This scheme is built with the MATLAB functions "constr" and "fmincon", which are classical enhanced numerical gradient-type optimizers. This gradient-based optimization scheme is applied to solve the developed model due to its efficiency and reliability in handling nonlinear problems. The process is iterative with the basic steps shown in the flowchart of Fig. 5. The optimization is initialized at a randomly chosen feasible starting point in order to enhance the chance of the algorithm to converge to global minima. The optimization algorithm makes use of both the model evaluation as well as the directed gradient to converge toward the optimal point [33, 34]. The error tolerance, ε, was set to be 1×10^{-5}.

2.2 Results and Discussion

The value of F, R_c, X_m, as well as other parameters of the system shown in the circuit of Fig. 2 vary with any variation in the controlled parameters (u, Z_L, and C). Under different load conditions, Fig. 6 presents the variations in X_m, R_c, V_o, and I_s, against C. The curve value of X_m drops to a minimum as C is being increased and then starts growing up as shown in Fig. 6a. On the other hand, the curve value of R_c decreases and increases irrespective of the value of X_m. Figure 6b shows that the stator current, I_s, increases as the excitation capacitance increases then it decreases against any increase in C. A similar observation is noticed for the terminal voltage, V_o in the same figure. It is worth mentioning that the machine operates as a generator when X_m is less than X_o (i.e., $X_m \leq X_o$) [1–4].

The variations in frequency, F, and minimum excitation capacitor (C_{min}) against the power factor (pf) at two different loads are shown in Fig. 7. This case was obtained by keeping the speed constant at 1.0 p.u. and X_m was assumed to have a value equal to X_o. The value of C_{min} is higher at the lower value of Z_L and vice versa. Opposite observation is noticed for F. In addition, as the power factor increases from zero to one, the frequency decreases in a small amount while the value of C_{min} increases in a small amount then decreases at a noticeable rate. When the capacitor and speed are fixed at 40 µF and 1.0 p.u., respectively, the variation in X_m and F versus power factor is shown in Fig. 8. It is noticeable that the larger the value of Z_L, the lower the value of X_m. Again, the frequency decreases in a small amount as the power factor increases from zero to one. The decrement is more for the minimum load impedance, Z_L. For the same value of the capacitor and speed of Fig. 8, the variation in the stator current and terminal voltage against the power factor is shown in Fig. 9. At a higher value of load impedance, the terminal voltage is approximately constant at a lower value of X_m as shown in Fig. 8.

The variation in R_c and X_m against u is shown in Fig. 10a for two different loads while C is kept fixed at 30 µF. For the same case, Fig. 10b shows the variation in the stator current and the terminal voltage versus speed. This figure shows clearly that R_c changes as the speed varies to agree to the measured results shown in Fig. 3.

Fig. 6 Variation against C for two different loads ($u =$ 1.0 p.u.): **a** Core loss resistance R_c and magnetizing reactance X_m, **b** stator current I_s and terminal voltage V_o

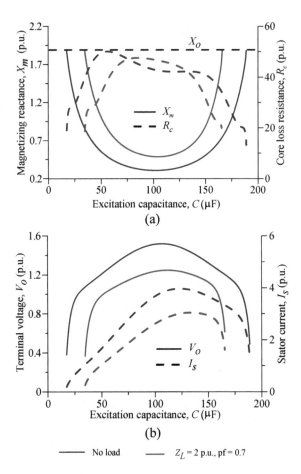

2.3 Experimental Verification

2.3.1 Setup

Under different conditions, experimental tests were carried out on the machine studied above to verify the proposed model. Figure 11 shows the experimental setup to obtain the performance of the machine under study. A DC motor that is controlled by a variable DC power supply drives the SEIG rotor. The excitation of the generator is achieved by using a star connected capacitor bank. The measurements of the mechanical and electrical quantities such as power, power factor, current, voltage, speed, and frequency were made by a computerized measurement unit (model CEM-U/Elettronica Veneta).

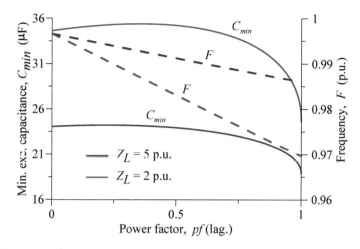

Fig. 7 Variation in frequency F and minimum excitation capacitance C_{min} against power factor pf for two different loads ($u = 1.0$ p.u.)

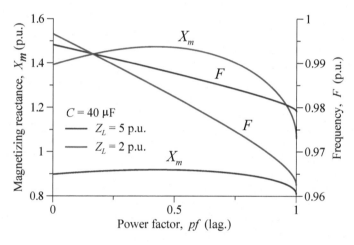

Fig. 8 Variation in frequency F and magnetizing reactance X_m against power factor pf for two different loads ($u = 1.0$ p.u.)

2.3.2 Performance Measurements

The variation in the stator current, I_s, and terminal voltage, V_o, versus C at a fixed speed ($u = 1.0$ p.u.) is shown in Fig. 12. The variation in I_s and V_o versus speed at a fixed capacitor ($C = 30$ μF) is shown in Fig. 13. Under different excitation capacitor values, Fig. 14 shows the system performance similar to that of Fig. 13. As expected, these figures show that an increase in speed or C will result in an increase in I_s and V_o. In addition, the frequency increases as u increases. These figures show

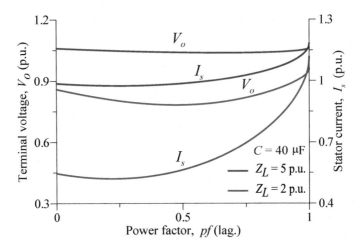

Fig. 9 Stator current I_s and Terminal voltage V_o against power factor pf for two different loads ($u = 1.0$ p.u.)

Fig. 10 Variation against speed for two different load conditions ($C = 30\,\mu F$): **a** Core loss resistance R_c and magnetizing reactance X_m, **b** stator current I_s and terminal voltage V_o

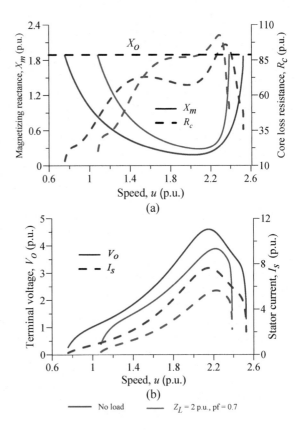

CMU: Computerized Measurement Units (Electrical Data (I, V, P, Q, Freq., pf,...), Mechanical Data (Speed)),
SEIG: Self-Excited Induction Generator, CB: 3-phase Capacitor Bank, PM: Prime Movers, SS: Speed Sensor

Fig. 11 Experimental setup for SEIG core loss study

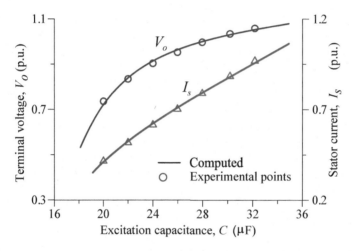

Fig. 12 Variation in stator current I_s and terminal voltage V_o against excitation capacitance C under no load ($u = 1$ p.u.)

good agreement between the simulated values and the measured results to validate the presented model.

2.3.3 Core Loss Influence

The effect of ignoring accurate modeling of core loss on the SEIG performance is discussed in this section.

The formulas that are used to evaluate the error in the values of efficiency (η) and terminal voltage (V_o) between the fixed value of R_c and the presented core loss

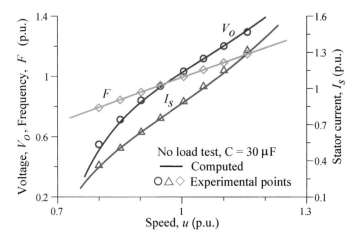

Fig. 13 Variation in stator current I_s, terminal voltage V_o, and frequency F against speed under no load

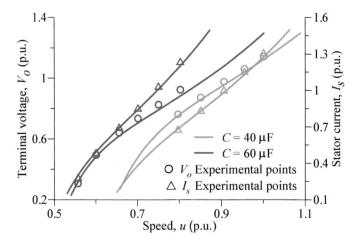

Fig. 14 Variation in stator current I_s and terminal voltage V_o against speed under no load

modeling are expressed by

$$\text{Error}(\eta) = \frac{\eta|_{R_c\text{isconstant}} - \eta|_{R_c=f(X_m)}}{\eta|_{R_c=f(X_m)}} \times 100 \tag{7}$$

$$\text{Error}(V_o) = \frac{V_o|_{R_c\text{isconstant}} - V_o|_{R_c=f(X_m)}}{V_o|_{R_c=f(X_m)}} \times 100 \tag{8}$$

Under different operating conditions, Figs. 15, 16, 17, 18, 19, 20 show the error evaluation for the system under study. The variation in the terminal voltage against

Fig. 15 Variation in terminal voltage V_o versus excitation capacitance C for two models of R_c

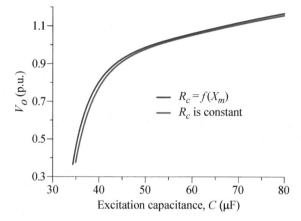

Fig. 16 Variation in the error of efficiency η and terminal voltage V_o against excitation capacitance C

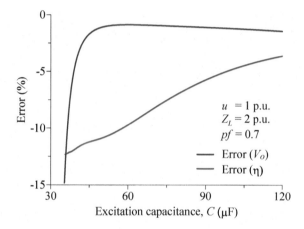

Fig. 17 Variation in terminal voltage V_o versus speed u for two models of R_c

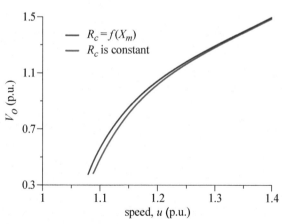

Fig. 18 Variation in the
error of efficiency η and
terminal voltage V_o against
speed u

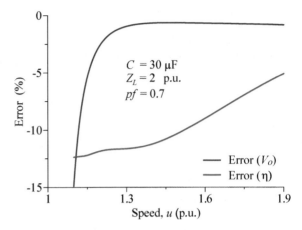

Fig. 19 Variation in the
error of terminal voltage V_o
against load impedance $|Z_L|$
for two models of R_c at
different power factors

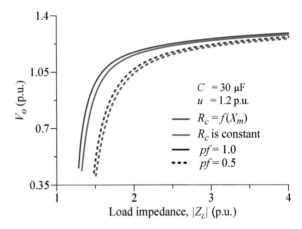

Fig. 20 Variation in error of
efficiency η and terminal
voltage V_o against load
impedance $|Z_L|$ at different
power factors

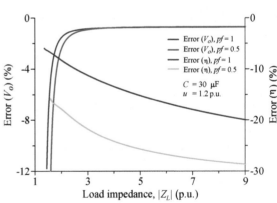

the excitation capacitor for two models of R_c is shown in Fig. 15. Figure 16 shows the error variation against excitation capacitance under fixed speed and load. The variation in the terminal voltage against the speed for the two models under fixed excitation capacitance and load is shown in Fig. 17 while the error variance of the same case is shown in Fig. 18. The conclusion from Figs. 16 and 18 is that at a low value of u and C, the error in the terminal voltage value is quite high and this error quickly decreases to an absolute low value as u, or C increase.

Figure 19 shows the variation in terminal voltage against load impedance for fixed and variable R_c under fixed excitation capacitance, power factor, and speed. Once again, at a low value of Z_L, the error in the terminal voltage value is quite high and this error rapidly decreases to an absolute low value as Z_L increases. On the other hand, the variation in the efficiency error grows up with the increase of load impedance.

3   Three-Phase Self-excited Reluctance Generator

This section presents more accurate modeling of core loss in SERG analysis taking into consideration the variation of saturation level when excitation capacitors, speed, or load change. To achieve this, the equivalent resistance of the core loss in the equivalent circuit of the generator is redeveloped, as presented below. Furthermore, a comparative study to assess the mentioned methods under different operation conditions and loads is presented. The effect of neglecting the core loss on the analysis of the SERG is discussed as well. The method presented in [35] is used as a reference to evaluate the errors in other methods.

3.1   Steady State Mathematical Model

Figure 21 shows the equivalent circuit of a three-phase self-excited reluctance generator under R–L load. For the performance at steady state, the following are assumed in this section to perform the analysis of the SERG [31]:

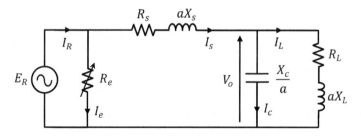

Fig. 21 Per phase equivalent circuit of the SERG

- Harmonic effects are neglected.
- The operation of the three-phase system is balanced.
- The core loss resistance R_e and the magnetizing reactance X_d are the only parameters that are affected by the saturation.
- The saturation in the quadrature axis is neglected.
- Only the reactances vary linearly with the frequency at variable speed operation.
- In the circuit of Fig. 21, all the parameters are assumed fixed and not affected by the saturation level except the magnetizing reactance X_d and the core loss resistance R_e.

The total impedance, Z_t, parallel to R_e in the circuit shown in Fig. 21 is given by

$$Z_t = R_t + jX_t = Z_s + (Z_L // Z_c) \tag{9}$$

where $Z_s = R_s + j\,a\,X_s$, $Z_L = R_L + j\,a\,X_L$, and $Z_c = -jX_c/a$.

Using d- and q-axis theory of salient pole synchronous machines, the currents and voltages are given as

$$\bar{I}_R = \bar{I}_d + \bar{I}_q \tag{10}$$

$$\bar{E}_R = -\bar{E}_q - \bar{E}_d \tag{11}$$

where $\bar{E}_q = j\,\bar{I}_d\,(aX_d)$, and $\bar{E}_d = j\,\bar{I}_q\,(aX_q)$.

In reluctance machine, the current must lead the air-gap voltage in order for it to operate as a generator. Hence, Eqs. (10) and (11) are represented as shown in Fig. 22 [30].

The current and voltage shown in Fig. 22 can be expressed as:

$$I_R^2 = I_d^2 + I_q^2 \tag{12}$$

$$E_R^2 = E_q^2 + E_d^2 = I_d^2(aX_d)^2 + I_q^2(aX_q)^2 \tag{13}$$

Fig. 22 Phasor diagram of the SERG

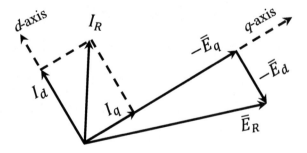

The reactive power is used as another link between the equivalent circuit of Fig. 21 and the phasor diagram of Fig. 22. The reactive power that is consumed by the total reactance in the circuit shown in Fig. 21, X_t, is equivalent to the generated air gap reactive power [31] and is expressed as

$$I_s^2(X_t) = I_q^2(aX_q) + I_d^2(aX_d) \tag{14}$$

Using Eq. (12) for I_R and applying the current divider rule to the circuit shown in Fig. 21 [31], I_e and I_s are derived as

$$I_e^2 = (I_d^2 + I_q^2)((R_t^2 + X_t^2)/H) \tag{15}$$

$$I_s^2 = (I_d^2 + I_q^2)(R_e^2/H) \tag{16}$$

where $H = (R_e + R_t)^2 + X_t^2$.

The air gap voltage E_R equals the voltage across R_e in the circuit shown in Fig. 21, and it can be expressed as

$$E_R = R_e I_e \tag{17}$$

Solving Eqs. (13), (14), (15), (16), and (17), and after some manipulation [31], the relationship between I_d and I_q is given by

$$I_d^2(a^2 X_d^2 H - R_e^2(R_t^2 + X_t^2)) = I_q^2(R_e^2(R_t^2 + X_t^2) - a^2 X_q^2 H) \tag{18}$$

$$I_d^2(aX_d H - R_e^2 X_t) = I_q^2(R_e^2 X_t - aX_q H) \tag{19}$$

By eliminating I_d and I_q from Eqs. (18) and (19), the following equation is obtained [31]:

$$\begin{aligned} X_d^2(aR_e^2 X_t - a^2 X_q H) &+ X_d(a^2 X_q^2 H - R_e^2(R_t^2 + X_t^2)) \\ &+ R_e^2 X_q((R_t^2 + X_t^2) - aX_q X_t) = 0 \end{aligned} \tag{20}$$

In cases where the core loss is neglected, Eq. (20) could be used by obtaining the limit as R_e goes to infinity, which yields the following equation:

$$X_d^2(aX_t - a^2 X_q) + X_d(a^2 X_q^2 - (R_t^2 + X_t^2)) + X_q((R_t^2 + X_t^2) - aX_q X_t) = 0 \tag{21}$$

It should be noted that the total impedance parameters R_t and X_t are functions of the variables a, R_s, X_s, X_c, and Z_L.

When considering core loss, Eq. (20) is used to obtain the performance of the SERG by solving for X_d, X_c, or a; otherwise, Eq. (21) is used in cases where the core loss is neglected.

3.1.1   Core Loss Modeling

Assuming the machine parameters R_s and X_s are fixed for a given generator, many variables are unknown in Eq. (20) such as a, X_d, R_e, X_c, and Z_L. Two of these unknowns, R_e and X_d, can be linked together by a function that experimentally can be determined and these two unknowns are considered as one unknown.

Practically, any change in speed, load, and/or excitation capacitance will cause a variation in the saturation level of the SERG and, consequently, cause a variation in the performance of the SERG and the core loss. This implies that R_e is not constant especially when considering variable speed [31]. Different methods that consider the core loss by adding resistance (R_e) to the model of the generator have been developed. The value of R_e was either fixed or proportional to the direct axis magnetizing reactance in a linear relation (i.e. $R_e = k\,X_d$) as proposed in [31]. More accurate modeling is proposed in this section by representing R_e as a polynomial function of X_d as

$$R_e = \sum_{i=0}^{n} m_i X_d^i \tag{22}$$

where n and m_i are the polynomial degree and coefficients, respectively, of the fitted curve obtained from the experimental results.

For a given machine, the parameters X_s and R_s are assumed to be fixed. Variables such as X_c, R_e, X_d, Z_L, and a are unknown in Eq. (20). In cases where R_e is represented as a linear or polynomial function of X_d, two of these unknowns (X_d and R_e) are considered as one unknown as they are linked together as mentioned above. The function coefficients k and m_i can be obtained from the linear and polynomial fitted curves of the experimental results of the core loss.

3.1.2   Method of Solution

Equation (20) can be solved for one unknown providing the other unknowns are given except in the case of X_d. When X_d is unknown, R_e is also unknown and is linked to X_d as given in the Appendix.

The method of solution used in this section involves the development of a program using Matlab that solves Eq. (20) directly. This program solves for X_d, X_c, a, or Z_L for minimum value of Eq. (20) (i.e. Eq. (20) = 0). Once the value of the unknown is obtained, the circuit of Fig. 21 is to be solved with the help of the magnetization curve to yield the performance of the generator.

With the use of the magnetization fitted curve, the performance of the SERG can be found just after solving Eq. (20) for the desired unknown. The basic circuit laws can be applied to the circuit shown in Fig. 21 to obtain electrical parameters such as voltages, currents, and powers.

3.1.3 Generator Performance

Once the unknown is obtained using Eqs. (20) or (21), The SERG performance can be evaluated as follows [31, 35]:

- For the given X_d, the value of I_d is calculated by using the polynomial function of the machine d-axis magnetizing fitted curve. i.e.

$$I_d = \sum_{i=0}^{n} k_i X_d^i \qquad (23)$$

where n and k_i are the polynomial degree and coefficients.

- Once I_d is obtained, either Eq. (18) or (19) can be used to calculate the value of I_q.
- Eqs. (12), (13), and (16) are used to evaluate I_R, E_R, and I_s, respectively.
- The terminal voltage V_o can be found as follows:

$$V_o = |E_R - I_s e^{j\theta_t} Z_s| \qquad (24)$$

where $\theta_t = tan^{-1}\left(\frac{X_t}{R_t}\right)$

- Load current, $I_L = V_o / \sqrt{R_L^2 + (aX_L)^2}$
- Capacitor current, $I_c = aV_o / X_c$
- Copper loss, $P_{cu} = I_s^2 R_s$
- Core loss, $P_{Fe} = E_R^2 / R_e$
- Output power, $P_o = I_L^2 R_L$
- Input power, $P_{in} = P_{cu} + P_{Fe} + P_o$
- Efficiency, $\eta = P_o / P_{in}$.

3.2 Results and Discussions

Machine data and the experimental results are given in [31] have been used in this section to verify the validity of the proposed model. The magnetization curve of the machine is shown in Fig. 23, to be used with the circuit of Fig. 21 to yield the SERG performance.

The variation of R_e against X_d according to the level of saturation is shown in Fig. 24. In addition, it can be seen from Figs. 23 and 24 that the saturation level in the generator is variable, and accordingly the values of X_d. and R_e.

Furthermore, Figs. 23 and 24 show the fitted polynomial functions of the core loss and magnetization curves. Third-degree polynomials were found to be good and suitable approximations for both R_e and I_d against X_d, and its coefficients are specified in the Appendix.

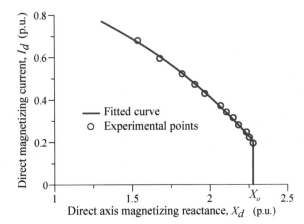

Fig. 23 Magnetization curve of the SERG under study

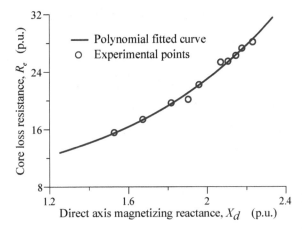

Fig. 24 Variation in R_e versus X_d of the SERG under study

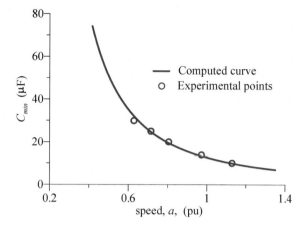

Fig. 25 Variation in C_{min} versus the speed under open circuit

The machine performance as a generator is computed theoretically by using the values obtained experimentally for X_o, and the curves of the magnetization and core loss. To make the machine works as a self-excited generator, a minimum capacitance (C_{min}) is required for excitation. Under open circuit, Fig. 25 shows the variations of C_{min} against the speed, a, and obviously it shows that C_{min} decreases as the speed of the generator increases. Additionally, the machine performance is tested under fixed speed for terminal voltage, V_o, variation against the excitation capacitance, C, as shown in Fig. 26. Furthermore and under fixed excitation capacitance, C, and speed, a variable resistive load is connected to the machine. The theoretically computed and measured variations of the terminal voltage, V_o, and the excitation capacitance current, I_c, against the load current, I_L, are shown in Fig. 27.

Figures 25, 26, and 27 show the variations of the computed and experimental results. These figures show the accuracy and superiority of the presented model, as can be seen from the perfect correlation between experimental and computed results.

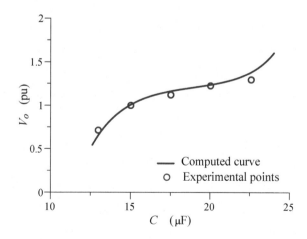

Fig. 26 Variation in V_o versus C under open circuit and fixed speed ($a = 1.0$p.u.)

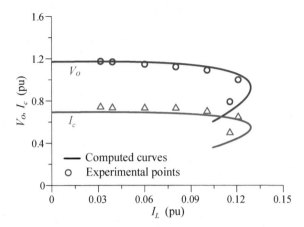

Fig. 27 Variation in V_o and I_c versus I_L of resistive load under fixed speed and excitation capacitance, C

3.3 Comparative Study of Core Loss Models in Three-Phase SERGs

As stated above, different methods that take the core loss into consideration by adding resistance (R_e) to the model of the generator have been developed. The value of R_e was either (i) fixed, (ii) proportional to the direct axis magnetizing reactance in a linear relation (i.e. $R_e = k\,X_d$), or (iii) a polynomial function of X_d, given by Eq. (22).

This section presents a comparative study to assess the above-mentioned methods under different operation conditions and loads. The effect of neglecting the core loss on the analysis of the SERG is discussed in this section. The method of representing the core loss as a polynomial function (i.e. Eq. (22)) is used as a reference to evaluate the errors in other methods [35].

Again, to obtain the performance of the SERG, Eqs. (20) or (21) need to be solved for one unknown in cases where the core loss is considered or neglected, respectively. In each consideration, all the other unknowns should be provided and given except in the cases (ii) and (iii) above when solving for X_d. In these two cases, when X_d is unknown, R_e is also unknown, but it is considered a function of the magnetizing reactance, X_d, as mentioned above.

A Matlab program was developed to solve Eq. (20) directly in the cases where the core loss is considered. A similar program was developed to solve Eq. (21) when the core loss is neglected. These two programs solve for a, Z_L, X_c, or X_d for minimum values of Eqs. (20) and (21), i.e., Eqs. (20) and (21) = 0.

Similar to Sect. 3.1.2, the performance of the SERG can be found just after solving Eqs. (20) or (21) for the desired unknown with the use of the magnetization fitted curve.

The performance of the SERG can be found just after solving Eqs. (20) or (21) for the desired unknown as given in Sect. 3.1.3.

Fig. 28 R_e versus X_d of the SERG under study

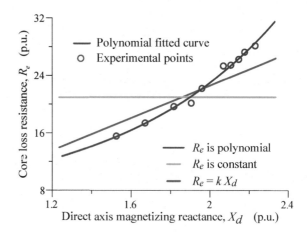

3.3.1 Results and Discussions

To perform the comparison of core loss models in three-phase SERGs, the machine data given in [31, 35] is used for this study. Figure 23 illustrates the d-axis magnetization curve of the machine that was used to obtain the performance of the SERG.

Figure 28 illustrates all models of R_e. In this study, the constant resistance was considered to be 21 p.u., which is equal to the resistance when the generator is working as a motor at rating voltage. When R_e is considered to be linearly proportional to X_d, the value of the factor k is obtained by taking a linear fitting curve of the experimental points of R_e in Fig. 28, and it is found to be equal to 11.3 [31].

The theoretical computations of the performance of the machine as a generator use the values that are obtained experimentally for the unsaturated magnetizing reactances X_o and the curves of the core loss and magnetization. A minimum excitation capacitance, C_{min}, is needed to excite the machine so that it can operate as a SERG. This minimum or critical value of the excitation capacitance varies with any change in the operation conditions [31, 35]. Under open circuit and load conditions, Fig. 29 depicts the error in the value of the minimum excitation capacitance, C_{min}, versus speed, a, while ignoring or considering the core loss with two models that are compared to the reference model (i.e. modeling core loss as a polynomial function). The formula used to evaluate the error in the value of C_{min} is given by

$$\text{Error}(C_{min}) = \frac{C_{min}|_{R_e=kX_d, R_e=\infty, \text{or constant}} - C_{min}|_{R_e=f(X_m)}}{C_{min}|_{R_e=f(X_m)}} \times 100 \quad (25)$$

It is obvious that the absolute error in the value of C_{min} is higher when the core loss is neglected. The error is reduced when R_e is represented by a constant value. Figure 29 shows that the error is a minimum when $R_e = k X_d$. Furthermore, the figure illustrates that the error in the value of C_{min} increases as the machine is loaded.

Fig. 29 Errors in C_{min} versus speed, a, for different loads ($X_d = X_o$)

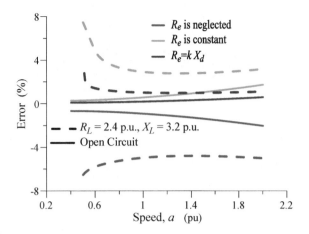

Furthermore, Fig. 30 shows the machine performance and the influence of changing the excitation capacitance, C, on the terminal voltage, V_o, under open circuit and load conditions at a constant rotor speed. The figure also shows that the operation range is narrowed when the machine is loaded. The formula used to evaluate the error in the value of V_o is given by

$$\text{Error}(V_o) = \frac{V_o|_{R_e = kX_d, R_e = \infty, \text{or constant}} - V_o|_{R_e = f(X_m)}}{V_o|_{R_e = f(X_m)}} \times 100 \qquad (26)$$

Again, Fig. 31 shows that the error is higher when the core loss is neglected and a minimum when $R_e = k X_d$. Furthermore, it also can be noticed from this figure that the error grows with the load at most of the operation points. Additionally, a variable resistive load is attached to the terminal of the machine under a fixed speed

Fig. 30 V_o versus C with different loads under constant speed ($a = 1.0$ p.u.)

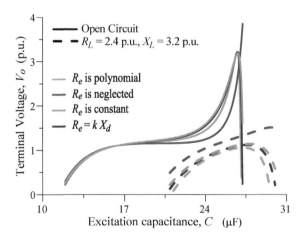

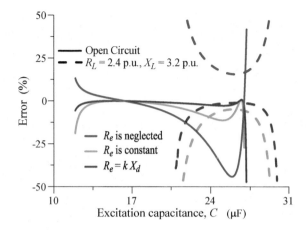

Fig. 31 Errors in the terminal voltage values (V_o) versus C with different loads under constant speed ($a = 1.0$ p.u.)

and excitation capacitance, C. Figure 32 shows the theoretically calculated variations in the excitation capacitance current, I_c, and the voltage, V_o, versus the load current, I_L. Under the same variable resistance, R_L, Fig. 33 shows the values of the load current, I_L, and the terminal voltage, V_o, versus the load resistance, R_L. The errors in the terminal voltage value versus the load resistance, R_L, is evaluated as given by Eq. (26) and shown in Fig. 34. Once again, the absolute error is a minimum when $R_e = k X_d$ and a maximum when the core loss is neglected. Figure 35 shows R_e versus C for different loads under a fixed speed. The graph of R_e versus R_L under constant speed and excitation capacitance is displayed in Fig. 36. Clearly, Figs. 35 and 36 show that the core loss varies with any change in the circuit parameters or operating conditions.

These figures show that the error of neglecting the core loss is relatively high which confirms the importance of including the core loss in the analysis of SERG.

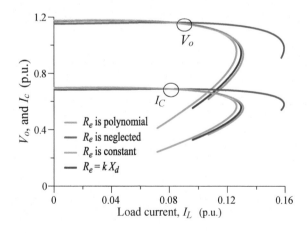

Fig. 32 V_o and I_c versus I_L of resistive load at fixed excitation capacitance and speed ($a = 1.0$ p.u., $C = 18$ μF)

Fig. 33 V_o and I_L versus R_L at fixed excitation capacitance and speed ($a = 1.0$ p.u., $C = 18$ μF)

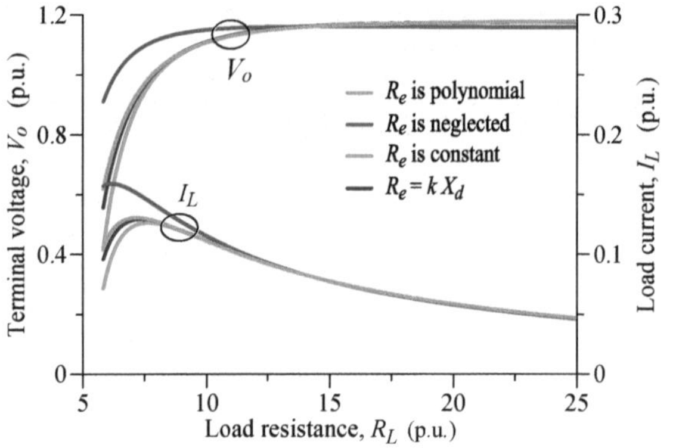

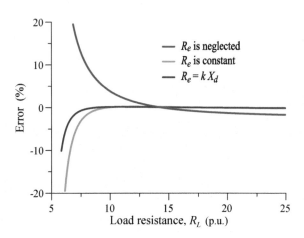

Fig. 34 Errors in the terminal voltage values (V_o) versus R_L at constant speed and excitation capacitance ($a = 1.0$ p.u., $C = 18$ μF)

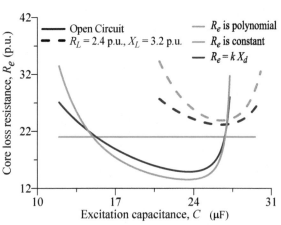

Fig. 35 R_e versus C for various loads at constant speed ($a = 1.0$ p.u.)

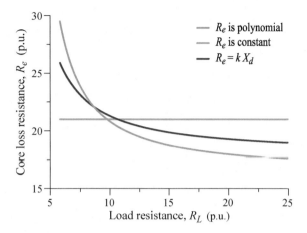

Fig. 36 R_e versus R_L at a fixed excitation capacitance and speed ($a = 1.0$ p.u., $C = 18\ \mu$F)

4 Conclusion

The performance of the Self-Excited Induction Generator (SEIG) has received excessive attention in the literature due to its advantages such as its brushless squirrel cage rotor construction, condensed size, absence of DC excitation, less maintenance cost, and superior transient performance. This makes the SEIG the simplest and cheapest energy conversion device suitable for Wind Energy Conversion Systems (WECS) in both grid and off-grid operations.

Self-excited Reluctance Generator (SERG), is another versatile AC generator that has also found applications in WECS. The SERG is a synchronous machine with a rotor designed with no windings and hence requires no field excitation [36]. Hence the SERG is also a self-excited generator with the advantages of being robust, having a simple rotor construction. Additionally, the SERGs are with low maintenance, cost, and losses [23]. The SERGs can also be operated in grid or isolated systems. In isolated systems, the excitation is achieved through the use of an appropriate capacitor bank connected across the stator terminals. In grid operation, excitation is attained from the grid [36].

This chapter introduces a novel method of modeling the core losses more accurately in SEIG analysis. The core loss is represented in the equivalent circuit by a variable resistance that is related to the level of saturations in the generator. Theoretical results were verified by comparing the outcomes with the experimental measurements. Also, detailed accurate modeling of core loss in SERG is developed in this chapter. To attain this goal, the equivalent resistance of the core loss in the circuit model is considered variable and also as a function of the saturation level in the magnetic circuit of the generator. Furthermore, this chapter presented a comparative study to assess the methods used to model the core loss in the SERG under different operation conditions and loads. The effect of neglecting the core loss on the analysis of the SERG was also discussed. From the study made, it is found out that the accurate model developed by this work showed that ignoring core loss leads to an error in the range of 2–12% in determining the terminal voltage and 10 -30% error in the value of efficiency. The usefulness of this accurate model is in its ability to present the terminal voltages of three-phase SEIG and SERG and their efficiencies very close to the actual practical values. Additionally, a comparative study of different approaches to modeling the core loss of SERG revealed that modeling R_e as a polynomial function of X_d as done by this work has overcome the errors of the previous work. The model made by this work showed that neglecting the core loss as done by some researchers leads to maximum error (in some applications) in determining the terminal voltage of the generator particularly at low voltage. Similarly, it showed minimum error for those who represent the core loss to be linearly proportional to the direct axis magnetizing reactance, X_d (i.e. $R_e = k\,X_d$). This study has added a degree of representing the SEIG and SERG more accurately by modeling the core loss resistance as a variable function related to the level of saturation. It should be mentioned that ignoring core loss should be carefully studied in some applications. In general, the error grows with the load at most of the operation points. In addition, changes in the machine parameters or operation conditions affect the saturation level and, accordingly, vary the core loss and the generator performance.

Appendix

Data of the Machines

The data of the machines used for the study of accurate modeling are given in Table 1.

Fitted Curves

The magnetizing curves given by Eqs. (1) and (23) can be fitted by two polynomials of third-degree. The coefficients of the polynomials are given in Table 2.

The core losses given by Eqs. (2) and (22) can be fitted by two polynomials of third-degree. The coefficients of the polynomials are given in Table 3.

Table 1 Specifications and p.u. parameters of the machines under study and base values

Quantity/parameter	SEIG	SERG
Type	Induction	Reluctance
Power (kW)	1.0	1.0
Phase voltage[a] (V)	220	220
phase current[a] (A)	2.9	2.5
Frequency[a] (Hz)	60	60
Synchronous speed[a] (rpm)	1800	1800
Base Impedance (Ω)	75.86	88
R_s	0.086	0.12
R_r	0.044	–
X_s	0.19	0.17
X_r	0.19	–
X_q	–	0.915
X_o	1.89	2.27

[a] Base values

Table 2 Coefficients of the polynomial functions of the air-gap voltage and magnetizing current fitted curves

Generator	k_0	k_1	k_2	k_3
Induction	1.1	−0.636	0.727	−0.321
Reluctance	1.2	−0.3983	0.1435	−0.0706

Table 3 Coefficients of the polynomial functions of core loss fitted curves

Generator	m_0	m_1	m_2	m_3
Induction	270.67	−472.71	303.76	−67.045
Reluctance	0	18.092	−11.35	4.03

References

1. Jabri AKA, Alolah AI (1990) Limits on the performance of the three-phase self-excited induction generators. IEEE Trans Energy Convers 5(2):350–356
2. Jabri AKA, Alolah AI (1990) Capacitance requirement for isolated self-excited induction generator. IEE Proc B Electr Power Appl 137(3):154–159
3. Alolah AL, Alkanhal MA (2000) Optimization-based steady state analysis of three phase self-excited induction generator (in English). IEEE Trans Energy Convers 15(1):61–65
4. Alnasir Z, Kazerani M (2013) An analytical literature review of stand-alone wind energy conversion systems from generator viewpoint. Renew Sustain Energy Rev 28:597–615
5. Sam KN, Kumaresan N, Gounden NA, Katyal R (2015) Analysis and control of wind-driven stand-alone doubly-fed induction generator with reactive power support from stator and rotor side. Wind Eng 39(1):97–112

6. Kheldoun A, Refoufi L, Khodja DE (2012) Analysis of the self-excited induction generator steady state performance using a new efficient algorithm. Electr Power Syst Res 86:61–67
7. Nigim K, Salama M, Kazerani M (2004) Identifying machine parameters influencing the operation of the self-excited induction generator. Electr Power Syst Res 69(2–3):123–128
8. Wang L, Lee C-H (1997) A novel analysis on the performance of an isolated self-excited induction generator. IEEE Trans Energy Convers 12(2):109–117
9. Kersting WH, Phillips WH (1997) Phase frame analysis of the effects of voltage unbalance on induction machines. IEEE Trans Ind Appl 33(2):415–420
10. Eltamaly AM (2002) New formula to determine the minimum capacitance required for self-excited induction generator. In: 2002 IEEE 33rd annual IEEE power electronics specialists conference. Proceedings (Cat. No.02CH37289), vol 1, pp 106–110
11. Malik NH, Haque SE (1986) Steady state analysis and performance of an isolated self-excited induction generator. IEEE Trans Energy Convers 1(3):134–140
12. Sharma A, Kaur G (2018) Assessment of capacitance for self-excited induction generator in sustaining constant air-gap voltage under variable speed and load. Energies 11(10):2509
13. Hashemnia M, Kashiha A (2012) A novel method for steady state analysis of the three phase seig taking core loss into account. In: The 4th Iranian conference on electrical and electronics engineering (ICEEE2012). Gonabad, Iran, pp 28–30
14. Farrag M, Putrus G (2014) Analysis of the dynamic performance of self-excited induction generators employed in renewable energy generation. Energies 7(1):278–294
15. Arjun M, Rao KU, Raju AB (2014) A novel simplified approach for evaluation of performance characteristics of SEIG. In: 2014 international conference on advances in energy conversion technologies (ICAECT), pp 31–36
16. Hamouda RM, Eltamaly AM, Alolah AI (2008) Transient performance of an isolated induction generator under different loading conditions. In: 2008 Australasian Universities power engineering conference. IEEE, pp 1–5
17. Haque MH (2009) A novel method of evaluating performance characteristics of a self-excited induction generator. IEEE Trans Energy Convers 24(2):358–365
18. Abu Elhaija W, Thalji J (2014) Self-excited reluctance generator: State of the art and future trends. Int J Energy Convers (IRECON) 12(4)
19. Wang Y-S, Wang L (2000) Steady-state performance of a self-excited reluctance generator under unbalanced excitation capacitors. In: 2000 IEEE power engineering society winter meeting. Conference proceedings, Singapore, Singapore, vol 1, pp 281–285
20. Alolah AI (1994) Static power conversion from three-phase self-excited induction and reluctance generators. Electr Power Syst Res 31(2):111–118
21. Singh B, Niwas R (2016) Performance of synchronous reluctance generator for DG set based standalone supply system. Electr Power Syst Res 133:93–103
22. Al-Salloum AM (1994) Operation of three phase reluctance motors fed from a single phase supply. Thesis, King Saud University, Riyadh, Saudi Arabia, MSc
23. Mohamadein AL, Rahim YHA, Al-khalaf AS (1990) Steady-state performance of self-excited reluctance generators. IEE Proc B Electr Power Appl 137(5):293–298
24. Abdel-Kader FE (1985) The reluctance machine as a self-excited reluctance generator. Electr Mach Power Syst 10(2–3):141–148
25. Rahim YHA, Mohamadien AL, Khalaf ASA (1990) Comparison between the steady-state performance of self-excited reluctance and induction generators. IEEE Trans Energy Convers 5(3):519–525
26. Lawrenson PJ, Gupta SK (1967) Developments in the performance and theory of segmental-rotor reluctance motors. Proc Inst Electr Eng 114(5):645–653
27. Alolah AI (1991) Capacitance requirements for three phase self-excited reluctance generators. IEE Proc C Gener Transm Distrib 138(3):193–198
28. Alolah AI (1992) Steady-state operating limits of three-phase self-excited reluctance generator. IEE Proc C Gener Transm Distrib 139(3):261–268
29. Fukao T, Yang Z, Matsui M (1992) Voltage control of super high-speed reluctance generator system with a PWM voltage source converter. IEEE Trans Ind Appl 28(4):880–886

30. Chan TF (1992) Steady-state analysis of a three-phase self-excited reluctance generator. IEEE Trans Energy Convers 7(1):223–230
31. Alolah AI (1995) Steady-state analysis of three-phase self-excited reluctance generators. Eur Trans Electr Power 5(2):133–138
32. Singh GK (2004) Self-excited induction generator research—a survey. Electr Power Syst Res 69(2–3):107–114
33. Xue N, Du W, Gupta A, Shyy W, Marie Sastry A, Martins JRRA (2013)Optimization of a single lithium-ion battery cell with a gradient-based algorithm. J Electrochem Soc 160(8):A1071–A1078
34. Kuhn HW, Tucker AW (1951) Nonlinear programming. In: Proceedings of the Second Berkeley symposium on mathematical statistics and probability. University of California Press, Berkeley, Calif , pp 481–492
35. Al-Senaidi SH, Alolah AI, Alkanhal MA (2018) More accurate modeling of core loss in self-excited reluctance generator. In: 2018 XIII international conference on electrical machines (ICEM). Alexandroupoli, Greece, pp 2264–2268
36. Rana S, Kar NC (2008)Steady-state analysis of self-excited synchronous reluctance generator. In: 2008 Canadian conference on electrical and computer engineering, pp 1617–1620

Different Approaches for Efficiency Optimization of DFIG Wind Power Generation Systems

Ahmed G. Abo-Khalil, Ali M. Eltamaly, and Khairy Sayed

Abstract The overall efficiency of the DFIG is superior when it is working close to the rated operating point and rated flux level. However, in light loads, optimal efficiency requires operation at a reduced flux level. In this chapter, several algorithms for increasing the steady-state efficiency that integrated with the wind power generation system are proposed. The proposed algorithms are based on the flux-Elevel reduction by calculating the optimum d-axis current and also by estimating the optimum reference rotor d-axis current by using Particle Swarm Optimization-Support Vector Regression (PSO-SVR) algorithm. The PSO is implemented to automatically perform the parameter selection in SVR modeling while the SVR is used to predict the optimum rotor d-axis current corresponding to the minimum total power loss. The input of the SVR is selected to be wind speeds, d-axis current, and generator power loss. The output of the SVR is the reference d-axis current. An experimental setup has been implemented in the laboratory to validate the theoretical development.

A. G. Abo-Khalil (✉)
Department of Electrical Engineering, College of Engineering, Majmaah University, Almajmaah 11952, Saudi Arabia
e-mail: a.abokhalil@mu.edu.sa

Department of Electrical Engineering, College of Engineering, Assuit University, Assuit 71515, Egypt

A. M. Eltamaly
Sustainable Energy Technologies Center, King Saud University, Riyadh 11421, Saudi Arabia
e-mail: eltamaly@ksu.edu.sa

Electrical Engineering Department, Mansoura University, Mansoura 35516, Egypt

K.A. CARE Energy Research and Innovation Center at Riyadh, Riyadh 11451, Saudi Arabia

K. Sayed
Electrical Engineering Department, Faculty of Engineering, Sohag University, Soha 82524, Egypt
e-mail: khairy_sayed@eng.sohag.edu.eg

1 Introduction

The increased energy consumption resulting from high industrial and population growth coupled with the issue of fossil fuel depletion has urgently demanded measures for future energy sources. Many compounds, including carbon dioxide, generated from power generation from fossil fuels pose a threat to the environment, and nuclear developments are not as reliable as they were in the aftermath of earthquakes in Japan. Paying attention to these problems, many research institutes are working to prepare measures at the global level, and the best solution is to introduce a development facility using renewable energy resources such as solar power, wind power, and wave power to replace the dependency on fossil energy sources. Despite comparatively high initial investments, the facilities using alternative energy are boldly under the government's initiative in many countries including developed countries in terms of fossil energy depletion and environmental pollution [1–3].

Among all renewables, wind energy is more economic and for this reason, it has been developed at a commercial level. However, the fixed- speed turbines are ineffective because it means that it is only possible to obtain the best power at a certain speed [4]. To solve this problem, the best solution is a variable speed wind energy development system that changes the speed of a turbine according to the external wind.

Currently, the variable speed turbines are the most used because they optimize the capture of wind and consequently the generation of energy in a wide range of variation of the wind speed. In case of very high wind speed, it is necessary to waste the excess of this one to avoid damages that compromise the physical integrity of the electromechanical conversion system. All turbines are, therefore, designed with some kind of control over the power to be converted. There are two types of control, they are stall control and pitch control, which control the movement around the longitudinal axis of the turbine blades. Among the existing concerns with the integrity of the electrical power system is frequency control. It is known that the frequency of the system is intrinsically linked to the flow of active power, and it is important to keep power generation in constant control. The system must have a power reserve so that it can provide frequency control.

In cases where the adopted configuration consists of using a doubly fed induction generator (DFIG), there is the advantage of being able to increase the injected power by controlling the electromagnetic torque that is generated. The stator winding of the DFIG is connected directly to the power system, while the rotor is connected to the system through a back-to-back converter. This method can supply the power to the system at a constant voltage and frequency even if the speed of the rotor changes with the wind speed [5, 6].

Recently, the use of DFIG has been well established, due to its capability to operate at a variable speed ranging from sub-synchronous speed to super-synchronous speed [7–9]. In the DFIG wind turbine system, as shown in Fig. 1, a back-to-back converter with about 30% of the total capacity exists between the grid and the rotor winding. The converter operates at the slip frequency of ±30% slip and the converter controls

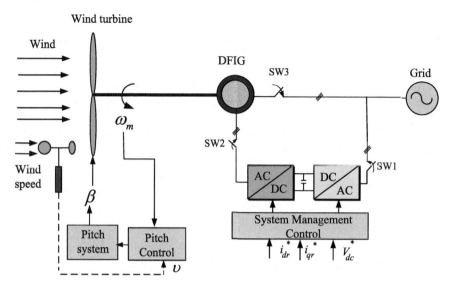

Fig. 1 The basic configuration of the DFIG wind turbine [19]

only the slip power. The converter excites the rotor side to control the wind turbine's output to generate and to control the reactive power and active power of the DFIG. In particular, DFIG's wind turbines are geared up to achieve maximum output [10–12].

Recently, improving the generator efficiency and minimizing the total losses getting more attention. The generator losses can be categorized as electrical copper losses, iron losses, mechanical and stray losses. The first two losses comprise about 80% of the total losses [13]. Generally, the generator flux is set to the rated value to have the best transient response and maximum torque capability. However, this condition increases the generator's total losses at low wind speed. When the generator runs at low speeds, the iron loss increases leading to a decrease in the total unbalance between copper and the iron loss. To control and minimize the total losses at low speeds, the generator flux should be optimized. The wind generator runs most of the time at low speed since the wind speed is lowered the rated value. In this case, the DFIG flux can be reduced by reducing the rotor d-axis current to reduce both copper and iron losses. Several loss minimization control schemes for induction machines using reduced flux levels have been reported in the literature. Many researchers have explored this principle and various methods have been proposed to obtain optimal control of the air-gap flux, especially for machines operating at light loads.

Several studies have presented different methods to improve the DFIG efficiency which can be classified into three categories. The first one is based on the DFIG model [14, 15], which determines the optimum rotor d-axis current by computing losses. The voltage, current, and flux equations are used to calculate the generator power and then obtaining the flux level which minimizes these losses. The advantages of this method are fast and smooth convergence. On the other hand, these methods are depending on the generator electrical parameters completely, so the operator needs accurate knowledge of the generator parameters to get the operating point correctly. Moreover, the operating conditions such as temperature variation and saturation may cause changes in the generator parameters, which may lead to an error in setting the optimum flux [16, 17]. The second category is based on relating the generator output power and d-axis by a given look-up table [18]. At any wind speed, the look-up table is used to determine the optimum d-axis current by interpolating the optimum efficiency operating point. This method needs a long time to search for the optimum flux level, which is not practical in continuously changing wind speed. The third category is the on-line methods, which depend on measuring the instantaneous torque and lowering the electrical output power settles in the maximum value for predetermined torque [19–22]. The advantage of this method is that there is no need to know the generator parameters since this method is based on searching the optimum flux level, and the insensitivity of this method to the generator parameter variations. However, this method is efficient when the generator runs at a constant speed which is not practical in wind power generation. By applying this method in variable wind power generation systems, it produces continuous flux and torque pulsations around the optimum operating point.

This chapter focuses on presenting the different loss minimization techniques on the DFIG system. The experimental setup is presented to validate the different techniques.

2   Losses in Wind Power Generation Systems

Since the Wind Energy Generation System (WEGS) consists of different mechanical components such as drive train gearbox, generator, etc., numerous losses can be found in the system. The total WEGS losses are divided into two main categories, mechanical and electrical losses. A general power flow diagram for a WEGS is shown in Fig. 2.

The gearbox losses are comprised of the gear mesh which depends on the instantaneous transmitted power and the no-load losses which consist of bearing, oil churning, and windage losses. The approximate percentage of the gear losses at rated load for typical large-scale wind turbines are listed in Table 1 [18].

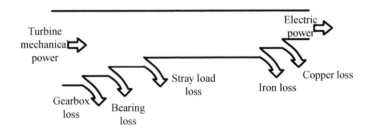

Fig. 2 Power flow in the wind power generation system

Table 1 Losses in the gear at rated load

Friction, windage, and oil churning losses	1.0% include turbine bearing (%)
Gear mesh losses	1.7
Total losses at rated load	2.7

Table 2 Losses in the generators at rated load

	fixed- speed WT	Variable-speed WT	
	IG (%)	Synch. generator (%)	PMSG (%)
Core loss	1.5	1.5	1.5
Cu. Losses, and additional losses	1.5 stator and rotor	1.15 stator	3.5 stator
Friction, windage and cooling losses	0.5	0.5	1.0
Excitation losses	–	0.75	–
Total	3.5	3.9	5.7

The generator losses can be divided into copper, hysteresis, eddy current core, windage, friction, and additional losses. The generator losses are a function of generator current, frequency, flux, and running speed. In large-scale wind turbines, the friction windage and cooling losses are high due to the bearing losses and the large cooling fan. The losses at the rated load of the different generators can be seen in Table 2 [18].

In order to propose a generator loss minimization algorithm in WECS, a relation between the generator losses and wind speed is necessary to be able to perfectly model these losses. It is known that the copper losses decrease in all the generators as the wind speed decreases. The core losses and friction losses are not reduced in the grid-connected induction generator because the flux linkage and the speed remain approximately constant. In the variable-speed systems, the windage and friction losses decrease when the generator speed decreases. The core losses of the directly driven permanent magnet generator do not decrease before the speed is decreased, since the flux linkage is constant. The core losses of the conventional synchronous generator decrease when the wind speed is below 12 m/s, approximate rated wind speed since the flux linkage is reduced by the excitation control. It is known that the generator copper losses decrease much faster than the other types of losses in all the generators. Therefore, the generator iron losses are more significant in the generator total losses than the copper losses in all generators which are used in variable-speed wind turbine systems.

In doubly fed induction generators, the voltage drop across the slip rings can be neglected due to the adjustment of the stator-to-rotor turns so that maximum rotor voltage is 75% of the rated grid voltage to have a safety margin. In Fig. 3, the DFIG losses are shown and it is clear that the generator losses are larger for high wind speeds for the variable speed induction generator (VSIG) system compared to the DFIG system. The reason is that the gearbox ratio is different between the

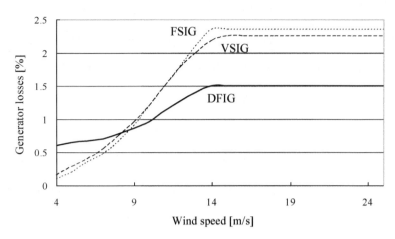

Fig. 3 Generator losses at different wind speed

two systems [23]. It can also be noted that the losses of the DFIG are higher than those of the VSIG for low wind speeds. The reason for this is that the flux level of the VSIG system has been optimized from an efficiency point of view while for the DFIG system the flux level is almost fixed to the stator voltage [24, 25]. This means that for the VSIG system a lower flux level is used for low wind speeds.

3   Losses in Doubly Fed Induction Generator

The generator losses consist of copper loss and iron loss, which are dependent on the current and flux level. At light load conditions, the iron loss can be decreased by reducing the flux level. The power converter losses depend on the current and switching frequency, which is of the lower portion of the total loss and is difficult to control.

3.1   Model-Based Loss Minimization Technique

The d-q equivalent circuits of DFIG considering the stator iron loss are shown in Fig. 4.

The generator copper losses can be expressed as shown in the following equation: [26]

$$P_{cu\_loss} = 1.5(i_{ds}^2 + i_{qs}^2)R_s + 1.5(i_{dr}^2 + i_{qr}^2)R_r \quad (1)$$

where

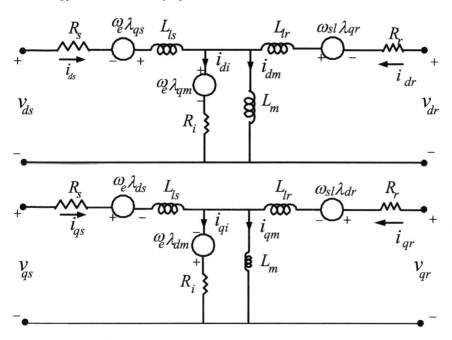

Fig. 4 Equivalent circuits of DFIG

i_{qs} Stator q-axis current [A];
i_{ds} Stator d-axis current [A];
R_s Stator resistance [Ω];
R_r Rotor resistance [Ω];
i_{dr} Rotor d-axis current;
i_{qr} Rotor q-axis current.

Substituting i_{dr} and i_{qr} in (1), copper loss can be expressed as shown in the following equation:

$$P_{cu\_loss} = 1.5\,[(R_s + (L_s/L_m)^2 R_r)i_{ds}^2 - 2\lambda_{ds} L_s R_r i_{ds}/L_m^2$$
$$((T_e/K_t)^2)R_s + (T_e L_s/K_t L_m)^2 R_r)\lambda_{ds}^{-2} \qquad (2)$$
$$+(\lambda_{ds}/L_m)^2 R_r)\,]$$

where
λ_{ds} Stator d-axis flux [Wb];
L_m Magnetizing inductance [H];
T_e Electromagnetic torque.
K_t Torque constant.

To minimize the total copper loss, the derivative of this loss with regard to i_{ds} should be obtained as shown in the following equation:

$$dP_{cu\_loss}/di_{ds} = 3R_s i_{ds} + 3\frac{R_r}{L_m^2}(-L_s\lambda_{ds} + L_s^2 i_{ds}) \tag{3}$$

To represent the DFIG iron loss, resistor R_i is used in the magnetic circuit as shown in Fig. 3 [27]. The stator flux linkages in the d-q reference frame are as

$$\lambda_{ds} = \lambda_{dm} + L_{ls}i_{ds} \tag{4}$$

$$\lambda_{qs} = \lambda_{qm} + L_{ls}i_{qs} \tag{5}$$

In rated torque, the flux is almost constant in d-q reference frame. However, the flux varies slowly according to speed variation when the generator works in the field weakening region [28]. Therefore, the flux variation can be neglected and the current flowing in the core loss branch can be expressed as shown in the following equations:

$$R_i i_{di} = -\omega_e \lambda_{qm} \tag{6}$$

$$R_i i_{qi} = \omega_e \lambda_{dm} \tag{7}$$

where R_i iron loss resistance.

Using (6) and (7), the current flowing in the core loss branch can be expressed as shown in the following equations:

$$i_{di} = -\omega_e(\lambda_{qs} - L_{ls}i_{qs})/R_i \tag{8}$$

$$i_{qi} = \omega_e(\lambda_{ds} - L_{ls}i_{ds})/R_i \tag{9}$$

The generator iron losses can be expressed as shown in the following equation:

$$P_{iron\_loss} = 1.5(i_{di}^2 + i_{qi}^2)R_i \tag{10}$$

Substituting (8) and (9) into (10) the generator iron losses can be expressed as shown in the following equation:

$$P_{iron\_loss} = 5[(L_{ls}i_{qs})^2 + \lambda_{ds}^2 - 2L_{ls}i_{ds}\lambda_{ds} + (L_{ls}i_{ds})^2]\omega_e^2/R_i \tag{11}$$

By substituting i_{qs} into (11)

$$P_{iron\_loss} = 1.5[(L_{ls}i_{ds})^2 + \lambda_{ds}^2 - 2L_{ls}i_{ds}\lambda_{ds} + (L_{ls}T_e/K_t)^2\lambda_{ds}^{-2}]\omega_e^2/R_i \tag{12}$$

With the same concept, the derivative of iron losses can be taken with regard to the d-axis current as shown in the following equation:

$$dP_{iron}/di_{ds} = 3\frac{\omega_e^2}{R_i}(L_{ls}^2 i_{ds} - L_{ls}\lambda_{ds}) \tag{13}$$

To minimize the total loss, the sum of the derivatives should be equal to zero as shown in the following equation:

$$d(P_{cu\_loss} + P_{iron\_loss})/di_{ds} = 0 \tag{14}$$

The derivative of the total losses can be expressed as shown in the following equation:

$$3\,[(R_s + (L_s/L_m)^2 R_r + \omega_e^2 L_{ls}^2/R_i)i_{ds} - (\omega_e^2 L_{ls}/R_i + L_s R_r/L_m^2)\,\lambda_{ds}\,] = 0. \tag{15}$$

After separating i_{ds} from the other variables, the stator d-axis current reference for minimum losses is given by

$$i_{ds}^* = \frac{(L_s R_r R_i + L_m^2 \omega_e^2 L_{ls})\lambda_{ds}}{L_m^2 R_s R_i + L_s^2 R_r R_i + L_m^2 \omega_e^2 L_{ls}^2} \tag{16}$$

The reactive power reference can be adjusted based on this value as shown in the following equation:

$$Q_s = \frac{3}{2}v_{qs}i_{ds}^* = \frac{3}{2} \cdot v_{qs} \cdot (v_{qs}/\omega_e - L_m i_{dr})/L_s \tag{17}$$

3.2 Loss Minimization Using PSO-SVR

In this method, instead of calculating the reference reactive power, a relation between the reference d-axis current, wind speed, and the minimum power loss in the DFIG are obtained using PSO-SVR. In this method, a prediction of unknown mapping between the inputs and output is conducted by selecting a series of training data. Wind speed, d-axis rotor current, and the total losses are employed as the input of the SVR, and the generator d-axis current reference samples were used as a target to train the SVR off-line. A fundamental interrelation between wind speed and generator d-axis current reference was predicted from the off-line training process. The predicted function and the instantaneous wind speed were then used on-line to determine the unknown d-axis current reference. When this mapping map is well constructed, the target output, which is the rotor d-axis current, can be obtained from the constructed relationship between the inputs, which are the wind speed, d-axis current, and power loss. For each sample, the measured wind speed, rotor d-axis current, and calculated DFIG minimum losses are used to be an input of the SVR function, and the predicted

Fig. 5 The structure of the optimum rotor d-axis current estimation

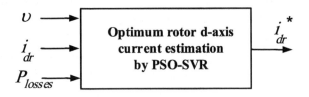

reference d-axis current for minimum losses is the output of the function as shown in Fig. 5.

The details of the proposed loss minimization control are discussed in the next section.

SVR is a regression method that considers the ε-insensitive loss function to predict random real values in the SVM (Support Vector Machine) used to predict the classification problem of training data. SVM is a recent learning method used for binary classification, and it is a technique of classifying data of two groups by obtaining a hyperplane that maximizes margin. Using a learning bias derived from statistical learning theory, a linear virtual space on a high-dimensional feature space that is trained by a learning algorithm will be used. This learning strategy is superior to other methodologies in a wide range of fields.

The purpose of the SVR is to find the hyperplane that minimizes the distance in all data. The regression function f (x) for predicting the optimal hyperplane targeted by the SVR is made as shown in Eq. (18) [28–30] as follows:

$$f(x) = (w^T . \varphi(x)) + b \tag{18}$$

where w^T is the weight vector and b is bias. $\varphi(x)$ is a space in which the vector x of the input space is nonlinearly mapped into a feature space of a high dimension. Including the ε-insensitive loss function and the slack variable ξ allows the optimization problem to be written as shown in Eq. (19) as follows:

$$Min \ \frac{1}{2}\|w\|^2 + \gamma \sum_{i=1}^{n} \Gamma(f(x_i) - y_i) \tag{19}$$

which is subjected to the following condition:

$$|y_i - w \cdot \Phi(x_i) - b| \leq \varepsilon + \xi_i$$
$$i = 1, 2, \dots\dots\dots n \quad \xi_i, \xi_i^* \geq 0 \tag{20}$$

where γ is a pre-specified value that controls the cost incurred by training errors. The slack variables, ξ_i and ξ_i^*, are introduced to accommodate the error on the input training set. Equation (19), $\frac{1}{2}\|w^2\|$ is a regularized term, ε is the permissible error, γ is a pre-specified value that controls the cost incurred by training errors, and the slack variables, ξ_i and ξ_i^*, are introduced to accommodate the error on the input training set.

Using the Lagrangian multiplier in Eq. (16), the general form of the SVR-based regression function is derived as follows [31]:

$$f(x) = \sum_{i=1}^{n} (\alpha_i - \alpha_i^*) \times K(x_i, x) + b \tag{21}$$

where α_i and α_i^* are Lagrange multipliers and $K(x_i, x_j) = \Phi(x_i)^T \Phi(x_j)$ is the kernel function.

$$K(x_i, x) = \exp\{-\frac{|x_i - x|^2}{\sigma^2}\} \tag{22}$$

In this study, radial base function (RBF) with [32] K is used as the kernel function, where σ is the RBF width. The parameters to be determined by the user in the SVR are the cost variable γ, the value of the ε-insensitive loss function, and the σ of the RBF. The optimal values of the three parameters were calculated through optimization using PSO.

PSO was first proposed by Kennedy and Eberhart (1995), and it is different from using natural selection evolution such as genetic algorithms. It has a parallelism feature and performs optimization using the concept of Swarm and Particle. Each randomly formed particle in the initial swarm travels the dimensional space by the number of parameters you choose to find the optimal value. After the initial Swarm search, the next Swarm remembers the value of the objective function for the location of each particle and shares the values with other particles to find the optimal solution. These features change from global search to regional search as generations pass and finally, converge to one point to find the optimal solution. In addition, the theory is simpler compared to other algorithms, making it easy to implement and efficient in operation.

In the PSO algorithm, the swarm has n agents specified particles, which is a set of user-selected parameter values. The equation of the PSO algorithm is as follows [33, 34]:

$$\begin{aligned}
v_{jk}(t+1) = {}& \omega_{damp} \bullet \omega(t) \bullet v(t) \\
& + c_1 \bullet r_1 \bullet (P_{best,jk}(t) - x_{jk}(t)) \\
& + c_2 \bullet r_2 \bullet (P_{best,k}(t) - x_{jk}(t))
\end{aligned} \tag{23}$$

where v_{jk} is the moving speed of parameter k of the jth particle in the group, ω_{damp} is the damping of the inertia control function, ω is inertia weight, c is acceleration constant, r is a random value, P_{best} is the optimal objective function value of the current swarm (particle), G_{best} is the optimal objective function value in all Swarm, and x_{jk} is the objective function value of the particle currently calculated. Particle velocity (v_{jk}) is the only operator of the PSO and is a multidimensional real vector representing the velocity of each parameter [35].

Particles in Swarm are used in Eq. (23) and it is generated depending on the speed of each parameter calculated in (23) and has dimensions as many as the number of parameters. Inertial load ω adjusts global search and local search capabilities when particles are formed. Initially, a large value is used to search globally, and as the number of iterations of the algorithm increases, the value of the inertia load decreases to allow a local search. The inertia control function ω_{damp} is a constant for convergence from wide-area search to local search by decreasing the value of inertia as the Swarm generation mentioned above. Acceleration constant c represents the stochastic acceleration of particles toward G_{best} and P_{best}. Smaller values of acceleration constants cause them to wander outside without going to the optimal solution, while higher values of acceleration constants can cause them to jump beyond the optimal solution or cause a sudden change of direction. The value can be between 0 and 1. In this study, the inertia load was set to 0.9, the acceleration constants c_1 and c_2 were set to 1.2, and the inertia control function was set to 0.9.

P_{best} and G_{best} are particles having the optimal values of the objective functions specified by the user among the particles in Swarm, which are the particle best and the global best, and P_{best} is the particle having the most optimal value among all the particles of Swarm. The final G_{best} of the algorithm thus represents the optimal solution of the given parameter. Since there is no G_{best} in the initial particle formation, particles are randomly formed only under the influence of P_{best}, but from the next generation of Swarm, particles are generated based on the previous G_{best} and P_{best} of Swarm.

The PSO algorithm adjusts its position according to the experience of neighboring entities. The entire multidimensional space formed by the parameters selected by the user can be explored. Particles formed in Swarm can be positioned by sharing the value and velocity of the objective function with other particles. Unlike the conventional methodology, the random probability optimization algorithm enables searching in uncertain domains and the calculation of operators is fast, so that an optimal solution can be easily produced. The relationship between global search and local search can be flexibly adjusted using inertial loads. If you are not familiar with model calibration, you can achieve significant accuracy if you only set the appropriate parameters and ranges. However, due to the stochastic approach, it is necessary to run the model more than a certain number of times in order to obtain good calibration results. The user's choice is based on the analysis results [36–38].

In this study, the PSO is implemented in a way to fast tune and optimize the SVR parameters. The framework of a PSO-SVR method is depicted in Fig. 6, which is described as shown in the following points:

(1) Collect training data (x_i, y_i) and determine the parameters ε, C, and σ using PSO.
(2) Determine the kernel function $K(x_i, y_i)$ from the collected data.
(3) Compute the Lagrange multipliers, b, and α_i by minimizing the quadratic function in

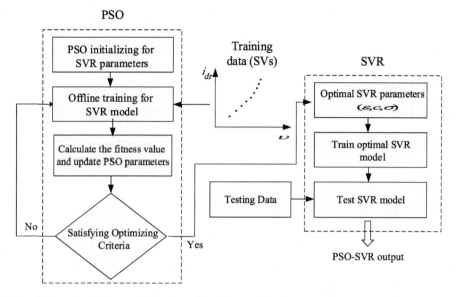

Fig. 6 Flowchart for loss minimization of DFIG using PSO-SVR

$$W(\alpha_i, \alpha_i^*) = \frac{1}{2} \sum_{i,j=1}^{n} (\alpha_i - \alpha_i^*)(\alpha_j - \alpha_j^*) K(x_i, x_j)$$

$$- \sum_{i=1}^{n} y_i(\alpha_i - \alpha_i^*) + \frac{1}{2C} \sum_{i=1}^{n} (\alpha_i^2 - \alpha_i^{*2})$$

(24)

subject to

$$\sum_{i=1}^{n} (\alpha_i - \alpha_i^*) = 0, \quad \alpha_i, \alpha_i^* \in [0, C]$$

$$b = mean\left(\sum_{i=1}^{n} \left\{ y_i - (\alpha_i - \alpha_i^*) K(x_i, x_j) \right\} \right)$$

(25)

All parameters in (25) are already off-line computed.

(4) Calculate the output of the estimator for any input.

Figure 7 shows the PSO-SVR training result for different wind speeds. The red-circled points are the training data which are included as support vector while the black cross points represent the data that are not supported vectors. The solid lines which go through the support vectors represent the learned model.

Figure 8 shows a block diagram of the rotor-side converter which controls the stator active and reactive power [36, 39]. The optimum output power P^* of the DFIG is used as the reference value for the power control loop. In this controller, the outer

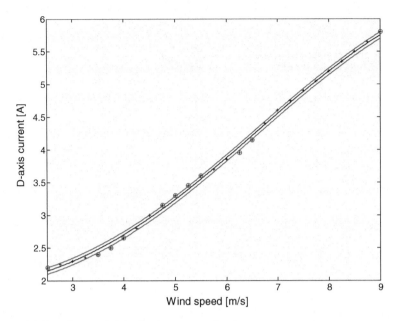

Fig. 7 RBF Kernel regression for d-axis current estimation

stator power control loop produces a reference value, i^*_{qr}, for the inner current control loop. The voltage command value generated in the process is implemented through a PWM signal.

The stator reactive power Q_s is controlled to the value to minimize the generator's total losses and to produce the d-axis rotor current reference.

Next, the grid-side converter is shown in Fig. 9. It plays a role in maintaining the DC-Link voltage of the system converter uniformly to transfer the power generated by the DFIG to the system without loss [27, 40–42].

4 Experimental Results

The wind power generator simulator, consisting of DFIG driven by a squirrel-cage induction motor as a wind simulator, was built in hardware and subjected to hardware experiments, as shown in Fig. 10.

To verify the validity of the suggested method, an experimental prototype was built and tested to control a 3-KW induction generator in a wind power generation system. The control algorithm was applied using a TMS320C33 DSP. Thy dynamic and steady-state characteristics of the induction generator with rated and reduced flux levels were tested. The parameters of the wind turbine and DFIG are listed in Tables 3 and 4 in the Appendix.

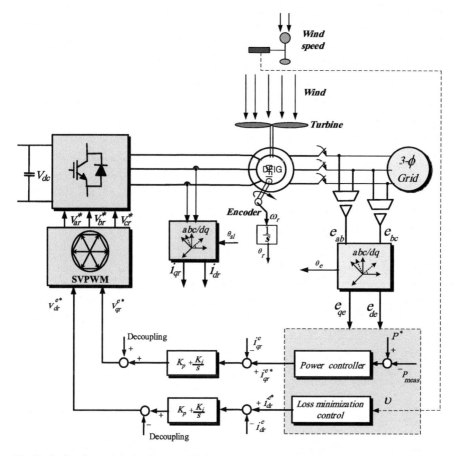

Fig. 8 On-line loss minimization control of rotor-side

Figure 11 gives the generator performance at 6 m/s wind speed. The DFIG is started with the rated flux current, and then the loss minimization algorithm is activated by estimating the optimum d-axis current and adjusting the reference reactive power to the value which reduces the total loss minimization to the minimum value. The calculated reference reactive power is shown in Fig. 11a. The loss minimization reactive power controller produces the optimum rotor d-axis current which decreased from the rated value to the minimum loss value. In this figure, the d-axis current is achieved its a steady-state reduced value of about 3.1A, as shown in Fig. 11b, very fast to achieve the minimum power loss. The stator d-axis current increased as a response of reactive current increasing as shown in Fig. 11c. The average power loss decreased from 128 to 71 W, which means about 43% can be saved. At the same time, the generator output power is increasing due to loss reduction as shown in Fig. 11d.

Figure 12 shows the generator performance at 7 m/s wind speed. The reduced d-axis current, in this case, is a little higher than the 6/s wind speed. The average d-axis

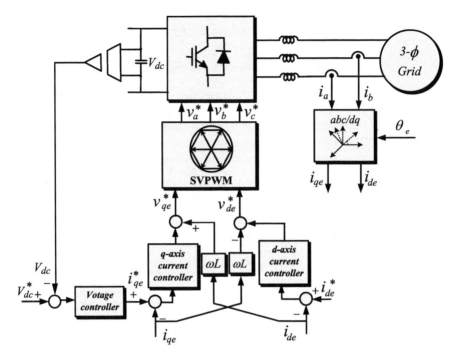

Fig. 9 DFIG grid-side control

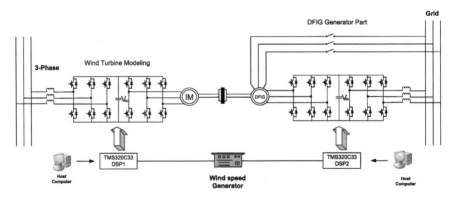

Fig. 10 Schematic of the experimental setup for a DFIG connected to the grid

current reduced steady state is about 3.5A as shown in Fig. 12b. The reduction in the d-axis current causes the increase in the reactive power and stator d-axis current of Fig. 12a and c. The average power loss is decreased also slightly from 150 to 90 W, which means about 40% can be saved in this condition.

As we can notice from Figs. 11 and 12, the power loss reduction is influenced mainly by the wind speed (and generator speed). The loss minimization process is

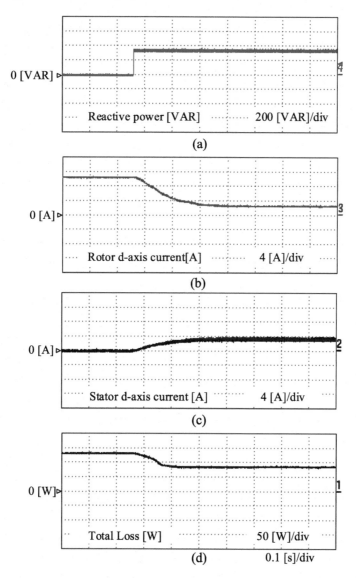

Fig. 11 Loss minimization at 6 m/s

more efficient at low wind speed and the percent of power saving in this rating can be more than 43%. Though this result is well known from the basic principles, it is confirmed that the proposed control algorithm is effective to reduce the induction generator operating loss.

Figure 13 shows the DFIG power loss when using conventional control and using loss minimization control. In very low wind speed, the power saving is higher when the loss minimization controller is activated. As the wind speed increases, the power

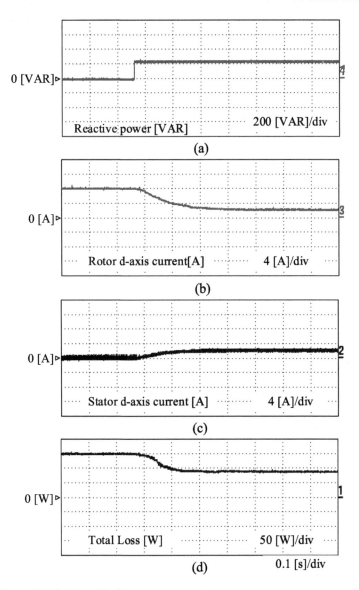

Fig. 12 Loss minimization at 7 m/s

saving decreases since the generator runs near the full load condition where the d-axis current is set to the rated value for high torque.

Fig. 13 Generator loss
minimization at low speed

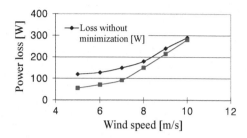

Table 3 Parameters 3[kW]
double-fed induction machine

Parameters	Value
Stator resistance	0.6 [Ω]
Rotor resistance	0.713 [Ω]
Iron loss resistance	155 [Ω]
Stator leakage inductance	0.003313 [H]
Rotor leakage inductance	0.066756 [H]
Mutual inductance	0.063443[H]

5   Conclusions

Wind power generation systems operate most of the time at a fraction of the rated
power due to the low wind speed. This operating condition causes a considerable
iron and copper loss, which reduces the total efficiency of the system. The DFIG is
usually controlled to extract the maximum power point by controlling the rotor q-
axis current while the reactive power is determined to control the generator excitation
and the desired grid power factor. The stator and rotor d-axis currents determine the
generator excitation level and the level of power loss.

In this chapter, a loss minimization control scheme for wind-driven DFIG was
proposed. The stator d-axis current is estimated to minimize the DFIG total losses by
controlling both stator's reactive power and rotor d-axis current wither in the constant
or variable wind/generator speed. A relationship between the rotor d-axis current,
wind speed, and power loss was deduced to calculate the optimum flux level which
minimizes the generator total losses. The experiment for 3[kW] has shown that it is
possible to reduce the power loss by up to 43% at 6[m/s] wind speed. Finally, this
method can be used in large-scale wind power generation systems to minimize the
total losses at low wind speed.

Table 4 Parameters of
turbine blade model

Parameters	Value
Blade radius	0.95 [m]
Max. power conv. coeff	0.45
Optimal tip-speed ratio	7
Cut-in speed	4 [m/s]

6 Appendices

The specification of the DFIG used for test is three-phase, four poles, 230[V], 60[Hz], 3[kW], of which parameters are listed in Table 3. The parameters of the wind turbine used are shown in Table 4.

References

1. Eltamaly AM (2007) Modeling of wind turbine driving permanent magnet generator with maximum power point tracking system. J King Saud Univ-Eng Sci 19(2):223–236
2. Abo-Khalil AG, Alghamdi A, Tlili I, Eltamaly A (2019) Current controller design for DFIG-based wind turbines using state feedback control. IET Renew Power Gener 13(11):1938–1949
3. Abo-Khalil AG (2012) Synchronization of DFIG output voltage to utility grid in wind power system. Elsevier J Renew Energy 44, 193–198
4. Park HG, Abo-Khalil AG, Lee DC, Son KM (2007) Torque ripple elimination for doubly-fed induction motors under unbalanced source voltage. In: Proceedings of the power electronics and drive systems PEDS,07, pp. 1301–1306
5. Abo-Khalil AG, Kim HG, Lee DC, Lee SH (2006) Grid connection of doubly-fed induction generators in wind energy conversion system. In: Proceeding of 5th international power electronics and motion control conference (IPEMC 2006). Shanghai, China, pp 14–16, 1–5
6. Zhong QH, Ruan Y, Zhao MH, Tan L (2013) Application of variable-step hill climbing searching in maximum power point tracking for DFIG wind power generation system. Power Syst Prot Control 41:67–73
7. Abrahamsen F, Blaabjerg F, Pedersen JK, Thogersen PB (2000) Efficiency optimized control of medium-size induction motor drives. Rome, Italy, IEEE-IAS Annu. Meeting, pp 1489–1496
8. Park HG, Abo-Khalil AG, Lee DC (2008) Wind turbine simulators considering turbine dynamic characteristics. Trans Korean Inst Electr Eng 57(4):617–624
9. Abo-Khalil AG (2015) Control system of DFIG for wind power generation systems. LAP LAMBERT Academic Publishing, ISBN-10: 3659649813, ISBN-13: 978-3659649813
10. Li Y, Xu Z, Zhang J, Wong KP (2018) Variable gain control scheme of DFIG-based wind farm for over-frequency support. Renew Energy 120:379–391
11. Khemiri N, Khedher A, Mimouni MF (2012) Wind energy conversion system using DFIG controlled by backstepping and sliding mode strategies. Int J Renew Energy Res 2:421–430
12. Abo-Khalil AG, Ab-Zied H (2012) Sensorless control for DFIG wind turbines based on support vector regression. In: Industrial Electronics Conference IECON. Canada
13. Abo-Khalil AG, Park H-G, Lee D-C, Ryu S-P, Lee S-H (2007) Loss minimization control for doubly-fed induction generators in variable speed wind turbines. In: ECON 2007-33rd annual conference of the IEEE industrial electronics society, pp 1109–1114
14. Abo-Khalil AG (2011) Model-based optimal efficiency control of induction generators for wind power systems. In: ICIT 2011, pp 191–197

15. Yang S-M, Lin F-C (2001) Loss-minimization control of vector-controlled induction motor drives. IEEE PEDS Conf Proc 1:182–187
16. Abo-Khalil AG, Kim HG, Lee DC, Lee S-H (2004) Maximum output power control of wind generation system considering loss minimization of machines. In: Proceedings of IECON'04, pp 1676–1681
17. Leidhold R, Garcia G, Valla MI (2002) Field-oriented controlled induction generator with loss minimization. IEEE Trans Ind Electron 49:147–156
18. Poitiers F, Bouaouiche T, Machmoum M (2009) Advanced control of a doubly-fed induction generator for wind energy conversion. Electr Power Syst Res 79:1085–1096
19. Lee D-C, Abo-Khalil AG Optimal efficiency control of induction generators in wind energy conversion systems using support vector regression. J Power Electron 8(4):345–353
20. Sousa GC, Bose BK, Cleland JG (1995) Fuzzy logic based on-line efficiency optimization control of an indirect vector-controlled induction motor drive. IEEE Trans Ind Electron 42:192–198
21. Chedid R, Mard F, Basma M (1999) Intelligent control of a class of wind energy conversion systems. IEEE Trans Energy Conv 14(4):1597–1604
22. Abo-Khalil AG, Lee DC (2008) Maximum power point tracking based on sensorless wind speed using support vector regression. In: IEEE Trans Ind Electron 55(3)
23. Grauers A (1996) Efficiency of three wind energy generator systems. IEEE Trans Energy Convers 11(3):650–657
24. Grauers A (1996) Design of direct-driven permanent-magnet generators for wind turbines. Ph.D. dissertation, Chalmers University of Technology, Goteborg, Sweden
25. Peterson A (2005) Analysis, modeling and control of doubly-fed induction generators for wind turbines. Ph.D. dissertation, Chalmers University of Technology, Goteborg, Sweden
26. Abo-Khalil AG, Lee DC, Seok JK (2004) Variable speed wind power generation system based on fuzzy logic control for maximum output power tracking. In Proceeding of power electronics specialists conference, vol 20–25. Aachen, Germany, pp 2039–2043
27. Eltamaly AM, Al-Saud MS, Abo-Khalil AG (2020) Dynamic control of a DFIG wind power generation system to mitigate unbalanced grid voltage. IEEE Access
28. Eltamaly AM, Al-Saud MS, Sayed K, Abo-Khalil AG (2020) Sensorless active and reactive control for DFIG wind turbines using opposition-based learning technique. Sustainability 12(9):3583
29. Eltamaly AM, Alolah AI, Abdel-Rahman MH (2010) Modified DFIG control strategy for wind energy applications. In: Proceeding of power electronics electrical drives automation and motion (SPEEDAM), vol 14–16. Pisa, Italy, pp 653–658
30. Abo-Khalil AG (2015) Control system of DFIG for wind power generation systems. LAP LAMBERT Academic Publishing: Latvia, ISBN-10: 3659649813, ISBN-13: 978-3659649813
31. Abo-Khalil AG, Alyami S, Sayed K, Alhejji A (2019) Dynamic modeling of wind turbines based on estimated wind speed under turbulent conditions. Energies 12(12):1907
32. YU BG, Abo-Khalil AG, Matsui M, Yu G (2009) Support vector regression based maximum power point tracking for PV grid-connected system. In: Photovoltaic specialists conference PVSC 34th. Philadelphia USA
33. Nuller KR, Smola A, Ratrch G, Scholkopf B, Kohlmargen J, Vapnik V (1997) Predicting time series with support vector machine in Pm/CA 1997. Springer LNCS 1327, pp 999–1004
34. Abo-Khalil AG, Lee D-C (2007) DC-Link capacitacne estimation using support vector regression in AC/DC/AC PWM converters. Korean Inst Electr Eng J 56(1):81–87
35. Abo-Khalil AG, Lee D-C (2008) DC-Link capaciance estimation in AC/DC/ACPWM converters using voltage injection. IEEE Trans Ind Appl 44(5):1631–1637
36. Lee K, El-Sharkawi M (2008) Modern heuristic optimization techniques. Wiley, Hoboken, N.J.
37. Eltamaly AM, Al-Saud MS, Abo-Khalil AG (2020) Performance improvement of PV systems, maximum power point tracker based on a scanning PSO particle strategy. Sustainability 12(3):1185
38. Abo-Khalil AG, Alyami S, Alhejji A, Awan AB (2019) Real-time reliability monitoring of DC-link capacitors in back-to-back converters. Energies 12(12):1907

39. Eltamaly A, Al-Saud M, AboKhalil AG, Farah H (2019) Photovoltaic maximum power point tracking under dynamic partial shading changes by novel adaptive particle swarm optimization strategy. Trans Inst Meas Control
40. Abokhalil AG (2019) Grid connection control of DFIG for variable speed wind turbines under turbulent conditions. Int J Renew Energy Res (IJRER) 9(3):1260–1271
41. Abo-Khalil AG, Alghamdi AS, Eltamaly AM, Al-Saud MS, Praveen RP, Sayed K, Bindu GR, Tlili I (2019) Design of state feedback current controller for fast synchronization of DFIG in wind power generation systems. Energies 12, 2427
42. Eltamaly AM, Alolah AI, Abdel-Rahman MH (2011) Improved simulation strategy for DFIG in wind energy applications. Int Rev Model Simul 4(2)

Voltage Source Converter Control Under Unbalanced Grid Voltage

Ahmed G. Abo-Khalil and Ali M. Eltamaly

Abstract The conventional control of the voltage source converter (VSC) assumes that the input voltage is balanced. However, unbalanced voltage is a phenomenon that occurs frequently in actual industrial sites. If the grid voltage is unbalanced, THD increases due to voltage negative component, and low harmonic components appear in DC-link voltage, which adversely affects the performance of the converter. Therefore, the purpose of this study is to propose an efficient control method that can solve the problem of the AC-DC converter due to the unbalance of grid voltage. A Multivariable State-Feedback (MSF) current controller is proposed to improve the performance of the VSC under grid voltage disturbances. The control process is carried out by adjusting the extracted positive and negative components of the grid d and q-axis currents. To minimize the DC-link voltage ripple, the reference negative grid currents are obtained from the DC-link voltage controller. However, if the target is to eliminate the imbalance of the grid current, the reference negative currents are set to zero. The experimental results are discussed to validate the proposed controller. The results show that the new MSF controller reduced the DC-link ripple and provides a fast dynamic response during unbalanced grid voltage.

Keywords Voltage source converter · Unbalanced grid voltage · MSF · PI controller

A. G. Abo-Khalil
Assiut University, Assiut, Egypt

School of Electrical and Computer Engineering, Yeungnam University, Gyeongsan, South Korea

A. M. Eltamaly (✉)
Mansoura University, Mansoura, Egypt
e-mail: a.abokhalil@mu.edu.sa; eltamaly@ksu.edu.sa

King Saud University, Riyadh, Saudi Arabia

© The Author(s), under exclusive license to Springer Nature Switzerland AG 2021
A. M. Eltamaly et al. (eds.), *Control and Operation of Grid-Connected Wind Energy Systems*, Green Energy and Technology,
https://doi.org/10.1007/978-3-030-64336-2_3

1 Introduction

The current electrical systems are dominated by large generating plants that are located at a great distance from the consumption centers and near the primary energy sources. This centralized model requires the construction of transmission lines to transport electrical energy and is characterized by having an acceptable cost–benefit ratio. However, the growing motivation to exploit resources locally, the tendency to stop relying on a single source of generation, and the need to reduce the emission of greenhouse gases (e.g., CO_2) into the atmosphere, creates the need to explore new alternatives to the centralized model [1].

Distributed generation (DG) emerges as a new option to the centralized model. This consists of the use of electric energy sources with powers and voltages below 10 MVA and 69 kV respectively, in order to generate energy locally and near the consumption centers. Generation from batteries, micro-turbines, solar panels, wind turbines, etc., are some of the clearest examples of this type of generation [2].

Loads using semiconductor power converters such as motor drive systems computer systems, etc., operate as nonlinear loads and are the main cause of generating serious harmonics on the power side. Such harmonic current can distort the power supply voltage and insulation breakdown or shortening of the lifespan of various power facilities such as electric devices, cables, and high-phase capacitors. Moreover, due to the nonlinear load, harmonic currents flow into the Point of Common Coupling (PCC) which makes the voltage distorted at the PCC and may cause the thermal overload of the power transformer. In addition, it causes malfunction or EMI phenomenon of various electronic equipment such as measuring equipment and causes electronic environmental problems such as interference in the communication system. In addition, abnormal vibration torque is generated in the motor, and losses such as iron loss and copper loss are increased [3, 4]. Recently, several types of active power filters are proposed [5] to minimize the harmonic distortion in the grid side and to compensate for the source voltage unbalance and sag. However, it is not an economical and effective method since the active power filter cannot be installed for every nonlinear load.

Advances in DG allow the use of new technologies that make the exploitation of electricity networks more flexible and secure. The integration of DG in electrical systems allows: Increase the degree of stability of the network, increase the levels of penetration of renewable energies, increase the limits of stability of the electrical system and improve the quality of energy [3].

Ideally, the voltages in the electrical systems are balanced, however, due to the existence of nonlinear and unbalanced loads, differences in the voltage levels of each of the phases of the network are presented.

Among the devices used to mitigate this type of disturbances is the voltage-controlled converter or (Voltage Source Converters) VSC whose connection is in parallel and its operation consists in general terms in the injection of compensation currents to mitigate disturbances related mainly to the currents demanded by disturbing loads. Articles in the technical literature and reporting control structures for power converters normally use PI controllers [6, 7]. These PI controllers require extensive knowledge of the system and must linearize and work at a point of operation. In fact, the converters are non-linear systems and, due to the change in the operating point, they may work improperly. The implementation of the control for electronic-based converters normally requires cascade controls with an internal current PI control loop and an external voltage PI control loop, this requires that the PI controllers be very well-tuned and that a relatively large capacitor is necessary to minimize the DC-link voltage ripple. To overcome this issue, nonlinear controllers such as back-stepping control [8], passivity-based control [9–11], and slid-mode control [12], direct power control [13] have been proposed by several researchers. Also, the input–output linearization based on feedback linearization has been applied for VSC control [14]. The main reason to use this control method is the ability to transform the nonlinear system into an uncoupled linear system. Then, a linear controller can be used to ensure stability throughout the entire operating range of the converter. It is interesting to note that when the DC voltage is selected as an output to be controlled in the exact linearization, the VSC is not completely linearizable and then a nonlinear internal dynamics appears that is unstable [15]. To overcome this inconvenience, internal current control loops are generally considered, where the exact linearization can be applied considering the currents as output. Then, a slower external DC voltage control loop is designed via a cascading PI control [16–18].

However, in actual three-phase power systems, the supply voltage is often distorted or unbalanced due to nonlinear loads and imbalance of steady-state loads which are connected to the point of common coupling (PCC). During unbalanced grid voltage operation, a negative component for grid voltage and current exists and produces oscillations of both active and reactive powers at the double- frequency of the grid. In this case, a dual current controller is proposed to separate the grid current into normal and reverse phases to eliminate the ripple of DC-link voltage [19, 20]. A method of using a proportional resonant current controller has been proposed to eliminate the effect of the unbalanced grid voltage in [21, 22]. However, using a dual current controller adds more complications for the controller design and gain calculations. Besides, this controller can compensate for the harmonic distortions in the resonance frequency only without compensating the other components. A method of predicting the future current and power components is introduced based on the predictive control method [23, 24].

2 The Requirement for Generation Interconnection Distributed to the Grid

The connection of DG to the power grid must meet a series of technical requirements to achieve efficient and safe operation in the presence of these. In particular, the voltages and currents in the network must be maintained at all times in their admissible ranges and the quality of service and power of the clients connected to the network in the area of influence of the generator must not be degraded. This translates on the one hand into operating conditions and restrictions that are imposed on the distributed generator to authorize it to inject power into the network, and on the other hand, the connection of the generator to the network is to assess the feasibility of certain proposed connectivity. Both concepts are denoted in the so-called "connection criteria" for distributed generation. These DG connection criteria are established based on the current electrical regulations applicable to the network planning and operation distribution system, and international standards of reference in the field. Among the latter, the "1547 IEEE Standard for Interconnecting Distributed Resources with Electric Power Systems" standard stands out.

The IEEE 1547 standard provides a set of requirements for the interconnection of distributed resources with EPS (electrical energy systems) for performance, operation, testing, safety, and maintenance of the interconnection.

Justification is made from the perspective of the distributor of some of the distributed generation connection criteria established in the IEEE 1547 standard, and their formulation and practical implementation are described and exemplified, among the most important connection criteria are:

A. Voltage Regulation at the Connection Point

The distributed generator is required not to actively regulate the voltage at the connection node. It is then considered that at the distribution level the generator connection node is a PQ (load) node and not a PV (generation) node as usual in generators connected to the transmission network. One of the justifications for this criterion is to minimize the chances that the distributed generator will be operating on an unintentional island. Another argument that supports the criterion is that there is no interference with the mechanisms and equipment of voltage regulation, for example, the case of current-controlled line voltage regulators. It is admitted that the distribution unit may require the generator to maintain the power factor measured at the point of coupling with the network within a given range (0.95–1.05), to help maintain the voltage profiles of the network within acceptable values by ± 5% without causing fluctuations.

B. Tensions in the Network under Normal Operating Regime

The criterion establishes that the presence of DG should not cause the voltage in the network nodes to exceed the regulatory limits for any state of network load and power injected by the generator. This criterion applies to the design of the connection of the generator to the electricity grid and for the operation of the network once the generator is connected, this makes the most restrictive case for the design and operation of the scenario of minimum load in the network, with the generator injecting all its authorized active power. Under these conditions, it must be ensured that the voltages in the network do not exceed the maximum permissible value, in the normal operating configuration. The range of admissible voltages varies according to the voltage level. A typical value is 5% around the nominal voltage value.

3 VSC Model During Unbalanced Grid Voltage

The voltage-controlled converters or VSC are one of the most used topologies in DG systems and high voltage transmission systems in direct current, with controlled magnitude and phase. This is achieved through the correct switching of the IGBTs, which are characterized by being efficient at the high switching frequency, reliable, and easily acquired in the market. Many of the control strategies used for these converters need to detect the fundamental component of the positive sequence of the network to determine the reference signals necessary to carry out the control.

These positive sequence detection methods are necessary to achieve the correct compensation between the GD-based generation systems and the power grid that are connected.

The apparent power of the VSC, in Fig. 1, during the grid unbalanced voltage can be written in terms of the positive and negative components of the grid voltage and current as [25]

$$S = P + jQ = E_{\alpha\beta} \cdot \overline{I_{\alpha\beta}} = \frac{\left[e^{(j\omega t)} \left(E_d^p + j E_q^p \right) + e^{(-j\omega t)} \left(E_d^n + j E_q^n \right) \right]}{\left[e^{(j\omega t)} \left(I_d^p + j I_q^p \right) + e^{(-j\omega t)} \left(I_d^n + j I_q^n \right) \right]} \quad (1)$$

where "p" and "n" are the subscripts of the positive and negative components, respectively. The apparent power in (1) is used to obtain the instantaneous active and reactive power $(p(t), Q(t))$.

In these conditions, the grid is composed of six components which are the average active and reactive power components (P_0, Q_0) and the four double-frequency components $(P_{c2}, P_{s2}, Q_{c2}, Q_{s2})$ as [26]

$$p(t) = P_0 + P_{c2} \cos(2\omega t) + P_{s2} \sin(2\omega t) \quad (2)$$

$$q(t) = Q_0 + Q_{c2} \cos(2\omega t) + Q_{s2} \sin(2\omega t) \quad (3)$$

where

Fig. 1 The voltage source converter circuit

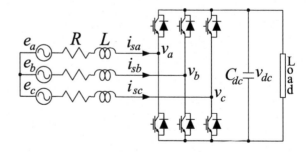

$$P_o = 1.5\left(E_d^p I_d^p + E_q^p I_q^p + E_d^n I_d^n + E_q^n I_q^n\right)$$

$$P_{c2} = 1.5\left(E_d^p I_d^n + E_q^p I_q^n + E_d^n I_d^p + E_q^n I_q^p\right)$$

$$P_{s2} = 1.5\left(E_d^p I_q^p - E_q^p I_d^n - E_d^n I_q^p + E_q^n I_d^p\right)$$

$$Q_o = 1.5\left(-E_d^p I_q^n + E_q^p I_d^p - E_d^n I_q^n + E_q^n I_d^n\right)$$

$$Q_{c2} = 1.5\left(-E_d^p I_q^n + E_q^p I_d^n - E_d^n I_q^p + E_q^n I_d^p\right)$$

$$Q_{s2} = 1.5\left(E_d^p I_d^n + E_q^p I_q^n - E_d^n I_d^p - E_q^n I_q^p\right)$$

E_{dq}^p and E_{dq}^n are the components of the grid d and q-axis voltages, respectively. The unbalanced three-phase grid voltage in dq-axis components are expressed as

$$E_{dq}^p = \left(E_d^p + j E_q^p\right) = \frac{2}{3}\left(E_a^p + a E_b^p + a^2 E_c^p\right) \tag{4}$$

$$E_{dq}^n = \left(E_d^n + j E_q^n\right) = \frac{2}{3}\left(E_a^n + a E_b^n + a^2 E_c^n\right) \tag{5}$$

where $a = e^{\left(j\frac{2\pi}{3}\right)}$, $E_a^p(t)$ is the positive component of $E_a(t)$, while $E_a^n(t)$ is a negative sequential component of $E_a(t)$.

The grid positive and negative currents I_{dq}^p, I_{dq}^n as well as the converter voltage, V_{dq}^p, V_{dq}^n can be defined in the same manner as E_{dq}^p and E_{dq}^n. The conventional electrical equations on the grid side of the VSC are stated as follows:

$$E_{dq}^p = V_{dq}^p + L\frac{d}{dt}I_{dq}^p + j\omega L I_{dq}^p + R I_{dq}^p \tag{6}$$

$$E_{dq}^n = V_{dq}^n + L\frac{d}{dt}I_{dq}^n + j\omega L I_{dq}^n + R I_{dq}^n \tag{7}$$

Using (2)–(7), the reference values for the d and q-axis currents are expressed as:

$$\begin{bmatrix} i_d^{p*} \\ i_q^{p*} \\ i_d^{n*} \\ i_q^{n*} \end{bmatrix} = \frac{2}{3D} V_{dc}^* I_{dc}^* \begin{bmatrix} e_d^p \\ e_q^p \\ -e_d^n \\ -e_q^n \end{bmatrix} \tag{8}$$

where

$$D = (e_d^p)^2 + (e_q^p)^2 - (e_d^n)^2 - (e_q^n)^2 \neq 0$$

where V_{dc}^* is reference DC-link voltage, I_{dc}^* the reference DC-link current reference which can be obtained from the DC-link voltage controller output. When the negative- currents components are adjusted as in (8), the DC-link voltage ripple can be controlled without eliminating the grid-side current unbalance. When the load is sensitive to voltage oscillation, the DC-link ripple illimitation is more important than regulating the grid currents to be balanced. However, to eliminate the unbalance in the grid currents the negative d and q-axis components in (8) should be adjusted to be zero [31–33].

4 Multivariable State Feedback Control

The VSC model in the rotating dq reference frame is given by [27]

$$V_d = E_d - R_g I_d - RL_g \frac{dI_d}{dt} + \omega_e L_g I_q \tag{9}$$

$$V_q = E_q - R_g I_q - RL_g \frac{dI_q}{dt} + \omega_e L_g I_d \tag{10}$$

$$P_{grid} = \frac{2}{3}\left(V_d I_d + V_q I_q\right) \tag{11}$$

$$C\frac{dV_{dc}}{dt} = I_{dc} - I_L = \frac{P_{grid}}{V_{dc}} - \frac{P_{Load}}{V_{dc}} \tag{12}$$

where Rg and Lg are the per-phase grid-side resistance and inductance of the coupling inductor, ω_e is grid angular frequency, I_L is the load current, and Co is the DC-link capacitance. P_{grid} and P_{Load} are the active power drawn from the grid and load.

The VSC state-space model for a time-invariant linear multivariable system can be represented in a synchronous frame as [28–30]:

$$\dot{x} = Ax + Bu + Ed \tag{13}$$

$$y = Cx \tag{14}$$

where, the state vector is x, the derivative of the space vector with respect to time is $\dot{x}$, the input or control vector is u, the system matrix is A, the input matrix is B, the disturbance matrix is E, the input disturbance vector is d, the output vector is y, and the output matrix is C.

$$x = \begin{bmatrix} i_{ds} \\ i_{qs} \end{bmatrix}, \quad u = \begin{bmatrix} v_{dr} \\ v_{qr} \end{bmatrix}, \quad d = \begin{bmatrix} e_{ds} \\ e_{qs} \end{bmatrix}$$

$$A = \begin{bmatrix} -\dfrac{R}{L} & \omega \\ -\omega & -\dfrac{R}{L} \end{bmatrix}, \quad B = \begin{bmatrix} -\dfrac{1}{L} & 0 \\ 0 & -\dfrac{1}{L} \end{bmatrix}$$

$$E = -B, \quad C = \begin{bmatrix} 1 & 0 \\ 0 & 1 \end{bmatrix}$$

and ω is grid angular frequency, the source currents are used as the state variables x, while the converter input dq- axis voltages are the input vectors u, the dq-axis grid voltage is the disturbance d, and grid current is the output y.

A. State Feedback Control

When $t \to \infty$, the control target is [38, 39]

$$\dot{x} \to 0 \ and \ y \to y_r$$

where y_r is a reference output.

The MSF controller has inaccurate steady-state performance due to model uncertainty and the type of this controller which is considered as a type of proportional control. Therefore, an integral function has to be added to minimize steady-state error. This function is expressed as:

$$p = \int_0^t (y - y_r)dt \tag{15}$$

Assuming that both d and reference y are constant, the derivative of (12) yield to:

$$\dot{p} = y - y_r = Cx - y_r \tag{16}$$

These equations can be written in matrix form and the state model can be expressed as

$$\begin{bmatrix} \dot{x} \\ \dot{p} \end{bmatrix} = \begin{bmatrix} A & 0 \\ C & 0 \end{bmatrix} \begin{bmatrix} x \\ p \end{bmatrix} + \begin{bmatrix} B \\ 0 \end{bmatrix} u + \begin{bmatrix} E & 0 \\ 0 & -I \end{bmatrix} \begin{bmatrix} d \\ y_r \end{bmatrix} \tag{17}$$

In a steady-state, both d and y_r are assumed to be constant which make $\dot{x}$ and error almost zero. Therefore, the state variables, proportional function, and the input vectors in steady-state must follow the following relationship:

$$\begin{bmatrix} E & 0 \\ 0 & -I \end{bmatrix} \begin{bmatrix} d \\ y_r \end{bmatrix} = -\begin{bmatrix} A & 0 \\ C & 0 \end{bmatrix} \begin{bmatrix} x_s \\ p_s \end{bmatrix} - \begin{bmatrix} B \\ 0 \end{bmatrix} u_s \qquad (18)$$

where the subscript "s" denotes a steady-state value,
Substituting (15) into (14) yields

$$\begin{bmatrix} \dot{x} \\ \dot{p} \end{bmatrix} = \begin{bmatrix} A & 0 \\ C & 0 \end{bmatrix} \begin{bmatrix} x - x_s \\ p - p_s \end{bmatrix} + \begin{bmatrix} B \\ 0 \end{bmatrix} (u - u_s) \qquad (19)$$

To represent the deviations in x_s, p_s, and u_s solutions, a new variable is proposed as:

$$z = \begin{bmatrix} z_1 \\ z_2 \end{bmatrix} = \begin{bmatrix} x - x_s \\ p - p_s \end{bmatrix} \quad (\dot{z} = \begin{bmatrix} \dot{x} \\ \dot{p} \end{bmatrix}) \qquad (20)$$

$$v = u - u_s \qquad (21)$$

Expressing (17) in the standard state space form as follows:

$$\dot{z} = \hat{A} z + \hat{B} x \qquad (22)$$

where

$$\hat{A} = \begin{bmatrix} A & 0 \\ C & 0 \end{bmatrix}, \hat{B} = \begin{bmatrix} B \\ 0 \end{bmatrix}$$

With applying linear state feedback control, the system in (19) is considered to be controllable and it can be written as:

$$\begin{aligned} v &= Kz \\ &= K_1 z_1 + K_2 z_2 \end{aligned} \qquad (23)$$

where,
K: feedback gain matrix;
K_1 and K_2: partitioned matrices.
By using Eqs. (16) and (17) the control function in (20) is expressed as

$$u = K_1 x + K_2 p = K_1 x + K_2 \int_0^t (y - y_r) dt \qquad (24)$$

B. Feedforward Control

It is expected to have zero static errors by applying the integral controller. This could be true in small disturbances, however, in large transients and disturbances the dynamic errors may be large. The dynamic performance can be improved during the large disturbances by adding a feedforward controller. The disturbance and reference inputs are both used to extract the feedforward control equations and the control system is defined as follows:

$$\begin{bmatrix} \dot{x} \\ \tilde{y} \end{bmatrix} = \begin{bmatrix} A & B \\ C & 0 \end{bmatrix} \begin{bmatrix} x \\ u \end{bmatrix} + \begin{bmatrix} E & 0 \\ 0 & -I \end{bmatrix} \begin{bmatrix} d \\ y_r \end{bmatrix} \tag{25}$$

where $\tilde{y} = y - y_r$,

The steady-state condition is reached when the left-hand side of (22) when the steady-state inputs and variable state are as follows:

$$\begin{bmatrix} x_s \\ u_s \end{bmatrix} = -\hat{G}^{-1}\hat{H} \begin{bmatrix} d \\ y \end{bmatrix} \tag{26}$$

where

$$\hat{G} = \begin{bmatrix} A & B \\ C & 0 \end{bmatrix}, \hat{H} = \begin{bmatrix} E & 0 \\ 0 & -1 \end{bmatrix}$$

The variables deviations from the steady-state give new variables which are expressed as:

$$\tilde{x} = x - x_s \quad (\dot{\tilde{x}} = \dot{x}), \tilde{u} = u - u_s \tag{27}$$

By substituting (24) into (24) the steady-state deviations are given by,

$$\dot{\tilde{x}} = A\tilde{x} + B\tilde{u}, \quad \tilde{y} = C\tilde{x} \tag{28}$$

By substituting (25) and (24) into (21), we get

$$u = K_1 x + [-K_1 \quad I] \begin{bmatrix} x_s \\ u_s \end{bmatrix} \tag{29}$$

Equation (22) is then used to define the feedforward by substituting into (26) as follows:

$$u = K_1 x + K_{ff} \begin{bmatrix} d \\ y_r \end{bmatrix} \tag{30}$$

Fig. 2 Block diagram for
MSF control

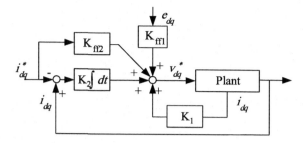

where k_{ff} is the feedforward gain and is defined as

$$K_{ff} = [K_1 \quad -I]\hat{G}^{-1}\hat{H}$$
$$= [K_{ff1} \quad K_{ff2}]$$

The control equation is then determined by substituting the integral controller (12) into (27) to be comprised from the state variables, reference inputs, and disturbance as follows:

$$u = K_1 x + K_2 \int_0^t (y - y_r)dt + K_{ff}\begin{bmatrix} d \\ y_r \end{bmatrix} \quad (31)$$

Figure 2 shows the MSF control block diagram including the feedback and feed-forward components. The detailed proposed control system is shown in Fig. 3 where the DC voltage V_{dc} is controlled to minimize the DC-link ripples which affect the connected load. A PI controller is used to control the DC voltage, while the current controllers for the positive and negative are MSF controllers as shown in the figure. The output reference DC-link current I_{dc}^* is used to calculate the converter reference power P^*. By using the measured and calculated voltages, currents, and power, Eq. (8) is then used to extract the current controller reference dq-axis currents. The state feedback matrix is then determined from the dq-axis currents and grid voltages to get the voltage reference which is then used as input to the modulator.

5 Experimental Results

Experiments are conducted to verify the validity of the proposed algorithm. The switching frequency of the IGBT PWM converter is 5 [kHz], and the sampling periods of voltage control and current control are 100usec. The three-phase input voltage is 100Vrms, the DC output voltage command is 150 [V], and the converter load is 50 [Ω].

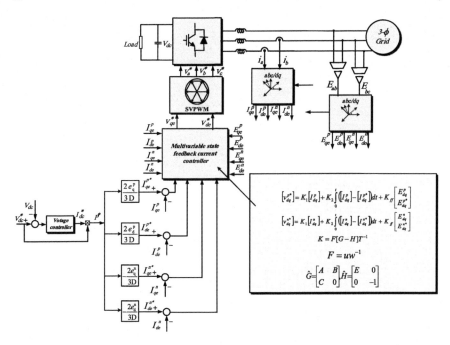

Fig. 3 Overall control block diagram of PWM converter

Figure 4 shows the waveform of the distorted and unbalanced grid voltage. Figure 5 shows the waveforms of a typical current controller under the conditions of Fig. 10. The input current in Fig. 5a contains harmonics and Fig. 5b shows that the DC-link voltage pulsates at twice the frequency of the power supply.

Next, Fig. 6 shows the waveform of adding both harmonic current controller and reverse-phase current controller. Figure 6a does not include harmonic currents, and Fig. 6b shows the waveforms when the DC-link voltage ripple adds almost all current controllers.

If the ripple of DC voltage due to unbalance does not greatly affect the control of the load or inverter connected to the DC terminal, it is recommended to set the reverse-phase current command to 0 to eliminate the unbalance of input current. Figure 7 shows the result of this control. Figure 7a shows that the input current is

Fig. 4 Unbalanced grid voltage

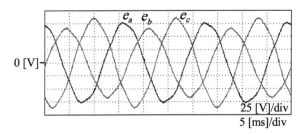

Fig. 5 Conventional current control at nonideal source voltage **a** three-phase input current **b** DC-link voltage

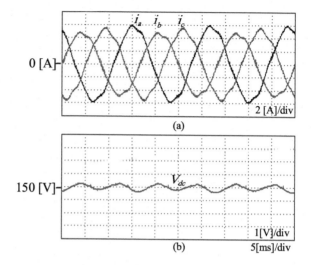

(a)

(b)

Fig. 6 Elimination of DC-link voltage ripples **a** three-phase input current **b** DC-link voltage

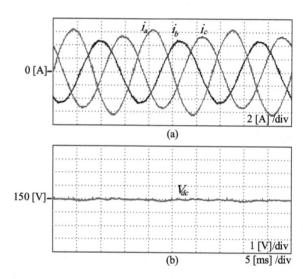

(a)

(b)

balanced, and Fig. 7b shows a slight increase in the ripple of the DC-link voltage in this case.

6 Conclusion

In this chapter, a control strategy for a three-phase PWM converter in case of unbalanced grid voltage has been proposed. The controller of the converter generally consists of a voltage controller in the outer loop and a current controller in the inner

Fig. 7 Grid current
balancing control without
DC-link voltage
a three-phase input current
b DC-link voltage

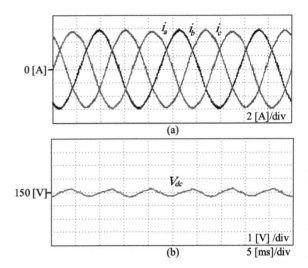

loop. The current controller in this study consists of four independent control loops: positive and negative- current controllers to maintain the input current balance and to eliminate the ripple of the DC terminal voltage appearing. The proposed controller exhibited excellent performance in minimizing the DC-link voltage ripple and in eliminating the unbalance of grid-side current.

References

1. Abo-Khalil AG, Kim HG, Lee DC, Seok JK (2004) Maximum output power control of wind generation system considering loss minimization of machines. Proc IECON'04, pp 1676–1681
2. Khaled U, Eltamaly AM, Beroual A (2017) Optimal power flow using particle swarm optimization of renewable hybrid distributed generation. Energies 10(7):1013
3. Qin H, Kimball JW (2014) Closed-loop control of DC-DC dual-active-bridge converters driving single-phase inverters. Power Electron IEEE Trans 29(2):1006–1017
4. Lin JL, Yao WK, Yang SP (2006) Analysis and design for a novel single-stage high power factor correction diagonal half-bridge forward AC-DC converter. Circ Syst IEEE Trans 53(10):2274–2286
5. Tsang KM, Chan WL (2005) Adaptive control of power factor correction converter using nonlinear system identification. Power Appl IEEE Proc Electr 152(3):626–633
6. Pena RS, Cardenas RJ, Clare JC, Asher GM (2001) Control strategies for voltage control of a boost type PWM converter. IEEE 32nd Power Elec. Spec. Conf, PESC'01, vol 2, pp. 730–735
7. Chang E-C, Liang T-J, Chen J-F, Chang F-J (2008) Real-time implementation of grey fuzzy terminal sliding mode control for PWM DC-AC converters. . IET Power Electron 1:235–244
8. Cecati C, Dell'Aquila A, Lecci A, Liserre M (2005) Implementation issues of a fuzzy logic-based three-phase active rectifier employing only voltage sensors. IEEE Trans Ind Electron 52:378–385
9. Allag A, Hammoudi M, Mimoune SM, Ayad MY, Becherif M, Miraoui A (2007) Tracking control via adaptive backstepping approach for a three phase PWM AC-DC converter. IEEE international symposium on industrial electronics, ISIE'07, pp 371–376

10. Escobar G, Chevreau D, Ortega R, Mendes E (2001) An adaptive passivity-based controller for a unity power factor rectifier. IEEE Trans Control Syst Technol 9:637–644
11. Harnefors L, Zhang L, Bongiorno M (2008) Frequency-domain passivity-based current controller design. IET Power Electron 1:455–465
12. Shtessel Y, Baev S, Biglari H (2008) Unity power factor control in three-phase AC/DC boost converter using sliding modes. IEEE Trans Ind Electron 55:3874–3882
13. Eltamaly AM, Alolah AI, Abdel-Rahman MH (2010) Modified DFIG control strategy for wind energy applications. In: SPEEDAM 2010. IEEE, pp 653–658
14. Mendalek N, Al-Haddad K, Fnaiech F, Dessaint LA (2003) Nonlinear control technique to enhance dynamic performance of a shunt active power _lter. IEE Proc Electric Power Appl 150:373–379
15. Lee T-S (2003) Input-output linearization and zero-dynamics control of three-phase AC/DC voltage-source converters. IEEE Trans Power Electron 18:11–22
16. Burgos RP, Wiechmann EP (2005) Extended voltage swell ride-through capability for PWM voltage-source rectifiers. IEEE Trans Ind Electron 52:1086–1098
17. Lee T-S, Tzeng K-S (2002) Input-output linearizing control with load estimator for three-phase AC/DC voltage-source converters. In: IEEE 33rd Power Electr. Spec. Conf., PESC'02, vol. 2, pp 791–795
18. Nikkhajoei H, Iravani R (2007) Dynamic model and control of AC/DC/AC voltage-sourced converter system for distributed resources. IEEE Trans Power Del 22:1169–1178
19. Savaghebi M, Jalilian A, Vasquez JC (2012) A secondary control level to focus the grid voltage and does not concern the local control objectives of the VSC. IEEE Trans Smart Grid 3(2):797–807
20. Eltamaly AM, Khan AA (2011) Investigation of DC link capacitor failures in DFIG based wind energy conversion system. Trends Electr Eng 1(1):12–21
21. Teodorescu R, Blaabjerg FM, Liserre M (2006) Proportional resonant controllers and filters for grid-connected voltage-source converters. IEE Proc-Electr Power Appl 153:750–762
22. Lascu C, Asiminoaei L, Boldea I, Blaabjerg FM (2007) High performance current controller for selective harmonic compensation in active power filters. IEEE Trans Power Electron 22:1826–1835
23. Dawei Z, Lie X, Williams BW (2010) Model-based predictive direct power control of doubly fed induction generators. IEEE Trans Power Electron 25:341–351
24. Cort'es P, Rodriguez J, Antoniewicz P, Kazmierkowski M (2008) Direct power control of an AFE using predictive control. IEEE Trans Power Electron 23:2516–2553
25. Ab-Khalil AG (2015) Control system of DFIG for Wind Power Generation Systems. LAP LAMBERT Academic Publishing, ISBN-10: 3659649813, ISBN-13: 978-3659649813
26. Abo-Khalil AG, Abdulbasser M (2016) Multivariable state feedback control of three-phase voltage source-PWM current regulator. Middle-East J Sci Res 24(3):10
27. Abo-Khalil AG, Ab-Zied H (2012) Sensorless control for DFIG wind turbines based on support vector regression. Industrial electronics conference IECON, Canada
28. Abokhalil AG (2019) Grid connection control of DFIG for variable speed wind turbines under turbulent conditions. Int J Renew Energy Res (IJRER) 9(3):1260–1271
29. Abo-Khalil AG, Alghamdi A, Tlili I, Eltamaly A (2019) A current controller design for DFIG-based wind turbines using state feedback control. IET Renew Power Generation 13(11):1938–1949
30. Abo-Khalil AG, Alghamdi AS, Eltamaly AM, Al-Saud MS, Praveen PR, Sayed K (2019) Design of state feedback current controller for fast synchronization of DFIG in wind power generation systems. Energies 12(12):2427

31. Abo-Elyousr FK, Youssef A (2018) Optimal PI microcontroller-based realization for technical trends of single-stage single-phase grid-tied PV. Eng Sci Technol Int J 21:945–956

32. Abdelwahab SAM, Yousef AM, Ebeed M, Abo-Elyousr FK, Elnozahy A, Mohammed M (2020) Optimization of PID controller for hybrid renewable energy system using adaptive sine cosine algorithm. Int J Renew Energy Res 10(2):669–677

33. Eltamaly AM, Alolah AI, Abdel-Rahman MH (2011) Improved simulation strategy for DFIG in wind energy applications. Int Rev Model Simul 4(2)

Robust Control Based on H∞ and Linear Quadratic Gaussian of Load Frequency Control of Power Systems Integrated with Wind Energy System

Ali M. Eltamaly, Ahmed A. Zaki Diab, and Ahmed G. Abo-Khalil

Abstract This chapter introduces a robust control scheme of load frequency control (LFC) of a micro-grid system integrated with a wind energy system. The control scheme is based on H∞ and linear quadratic Gaussian techniques. The main idea of the control design is to be stable against the parameter's uncertainties and the load disturbance. A complete model of the power system with the DFIG has been linearized. This model has been utilized to design both controllers of H∞ and linear quadratic Gaussian. The H∞ is designed by optimal selecting of the weighting functions to ensure the robustness and to enhance the overall performance. Also, the full states considering the frequency deviation are assessed based on the standard Kalman filter method. Moreover, the states of the system are applied with the linear quadratic Gaussian feedback optimal control performance under normal and abnormal operating conditions. Simulation tests are applied with the purpose of validation of the overall controllers' performance. The results proved the superiority of the planned integration of the wind energy system with the micro-grid.

Keywords Robust control · Load frequency control · Linear quadratic gaussian · H∞ · Wind energy system

A. M. Eltamaly (✉)
Sustainable Energy Technologies Center, King Saud University, Riyadh 11421, Saudi Arabia
e-mail: eltamaly@ksu.edu.sa

Electrical Engineering Department, Mansoura University, Mansoura, Egypt

K.A. CARE Energy Research and Innovation Center, Riyadh 11421, Saudi Arabia

A. A. Zaki Diab
Electrical Engineering Department, Minia University, Minia, Egypt
e-mail: a.diab@mu.edu.eg

A. G. Abo-Khalil
Department of Electrical Engineering, College of Engineering, Majmaah University, Almajmaah 11952, Saudi Arabia
e-mail: a.abokhalil@mu.edu.sa

Department of Electrical Engineering, College of Engineering, Assuit University, Assuit 71515, Egypt

© The Author(s), under exclusive license to Springer Nature Switzerland AG 2021
A. M. Eltamaly et al. (eds.), *Control and Operation of Grid-Connected Wind Energy Systems*, Green Energy and Technology,
https://doi.org/10.1007/978-3-030-64336-2_4

1 Introduction

In recent years, wind energy systems have been used to feed loads in remote and central areas. The wind energy applications are showing great interest due to their clean and sustainable nature. The advanced control system remedies the effects of the intermittent nature of the wind energy and it can be used widely even with weak power systems. This chapter is introduced to present a novel control system that enhances the stability of wind energy systems in case of normal and abnormal operation in micro-grids. In recent years, many control approaches are introduced to act with the impacts of load disturbance to regulate the frequency and voltages to the normal operating ranges. The variation in the frequency of the power system mainly depends on active power. Also, voltage control is affected by the variation of the reactive power too [1–4]. Therefore, the control problem of the power system is decoupled into two independent issues. The first one is the load frequency control (LFC), which is affected substantially by the active power control. The other one is to control the voltage based on regulating the reactive power. LFC is the main point of research in this chapter.

Energy and environmental topics are directed to integrating more renewable energies into power systems. Wind energy systems are rapidly increasing and its worldwide production is forecasted to produce to around 2000 GW in 2030 and 5000 GW in 2050 [5]. Hence, the integrations of the Wind Energy System (WES) with the power systems are very interesting topics for researchers. Many wind turbines (WTs) have been developed, among these WTs, the variable-speed wind turbines (VSWTs) have been considered as the best option for wind energy systems [6–12]. The inertial performance of WTs has been reported in Refs. [6, 10]. A comparison among the fixed-speed wind turbines (FSWTs) and doubly fed induction generator (DFIG) based WTs has been presented in [6]. Moreover, the simulation results had shown that FSWTs and DFIG-based WTs affect the frequency response of small power systems or micro-grids [6]. In Ref. [10], the presented results showed that an additional loop should be applied to enhance the machine inertial response of the WTs.

In recent years, many control schemes are applied to enhance the overall performance of such systems [12–14]. In Refs. [3, 15, 16], researchers applied the robust adaptive methods to act with the uncertainties of the parameters. Other researchers applied the Fuzzy logic controllers LFC in a two-area power system [17]. Others applied the artificial intelligence (AI) and genetic algorithms (GA) based controllers to LFC in [18, 19]. In Ref. [20], the H_∞ controller is applied and verified to control the photovoltaic pumping system. In Ref. [4], a robust control system concerning the linear quadratic Gaussian (LQG) and coefficient diagram method has been designed to LFC of a three-area power system. Also, adaptive model predictive control (AMPC) has been applied for solving the control issue of the LFC in the presence of a wind energy system [9]. For this reason, in this Chapter, the LFC for a single-area power system integrated with WES is considered using the combination of H_∞ and LQG controllers. The design methodology computes the optimal controller output considering constraints of the frequency deviation and load disturbance. The overall dynamic performance of H_∞ + LQG is simulated considering different operation conditions. A comparison among the system with H∞ alone, H∞ + WT, and H∞ + LQG + WT control scheme has been performed and presented in the results shown in this chapter. The simulation results proved the influence of the planned scheme. These results showed the superiority of the control system compared to many controllers of wind turbines [21–29].

The rest sections of this chapter have been ordered as follows: The mathematical model of the system has been presented in Sect. 2. The design of H_∞ has been introduced in Sect. 3. LQG is introduced in Sect. 4. The implementation of the overall control system has been presented in Sect. 5. Simulation results and discussions have been introduced in Sect. 6. The conclusion is introduced in Sect. 7.

2 System Dynamics

A. Mathematical description of a single-area system

The structured frequency response of a single-area power system is presented in Ref. [14]. The overall generator–load dynamic mathematical description is rewritten as shown in the following equations:

$$p.\Delta f = \left(\frac{1}{2H}\right) \cdot \Delta P_m - \left(\frac{1}{2H}\right) \cdot \Delta P_L - \left(\frac{D'}{2H}\right) \cdot \Delta f \tag{1}$$

$$p \cdot \Delta P_m = \left(\frac{1}{T_t}\right) \cdot \Delta P_g - \left(\frac{1}{T_t}\right) \cdot \Delta P_m \tag{2}$$

$$p \cdot \Delta P_g = \left(\frac{1}{T_g}\right) \cdot \Delta P_c - \left(\frac{1}{R.T_g}\right) \cdot \Delta f - \left(\frac{1}{T_g}\right) \cdot \Delta P_g \tag{3}$$

The diagram showing the logic used in this controller for the power system has been shown in Fig. 1. Equations (1–3) can be rewritten as a state-space model as shown in the following equations:

$$
\begin{bmatrix} p \cdot \Delta P_g \\ p \cdot \Delta P_m \\ p \cdot \Delta f \end{bmatrix} = \begin{bmatrix} -\frac{1}{T_g} & 0 & -\frac{1}{R.T_g} \\ \frac{1}{T_t} & -\frac{1}{T_t} & 0 \\ 0 & \frac{1}{2H} & -\frac{D'}{2H} \end{bmatrix} \begin{bmatrix} \Delta P_g \\ \Delta P_m \\ \Delta f \end{bmatrix} + \begin{bmatrix} 0 \\ 0 \\ -\frac{1}{2H} \end{bmatrix} \Delta P_L + \begin{bmatrix} \frac{1}{T_g} \\ 0 \\ 0 \end{bmatrix} \Delta P_c \tag{4}
$$

$$
y = \begin{bmatrix} 0 & 0 & 1 \end{bmatrix} \begin{bmatrix} \Delta P_g \\ \Delta P_m \\ \Delta f \end{bmatrix} \tag{5}
$$

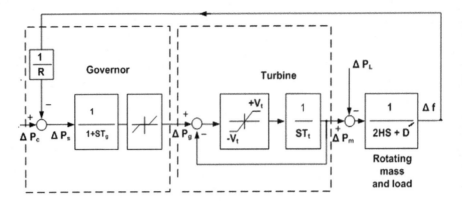

Fig. 1 The uncontrolled single-area power system

where ΔP_g denotes the governor's output variation. ΔP_m denotes the mechanical power variation. Δf denotes the frequency aberration. ΔP_L represents the load change while ΔP_c denotes supplementary control action; and p is a differential operator. While $\left(\frac{1}{s}\right)$ denotes integral Laplace. Furthermore, T_g and T_t represents the governor and turbine time constants. y denotes output. While H denotes an equivalent inertia coefficient. Also, D denotes equivalent damping constant. Moreover, R denotes the speed droop characteristic.

B. Mathematical Model of Wind Turbine based DFIG

Figure 2 displays a model of DFIG WT using the frequency response [14]. The presented model of the frequency response can be written as shown in the following

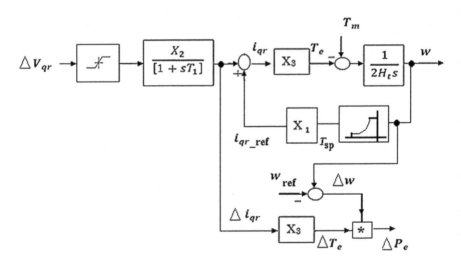

Fig. 2 Model of DFIG WT [14]

Table 1 The parameters shown in of Fig. 2	X_2	X_3	T_1
	$\frac{1}{R_s}$	$\frac{L_m}{L_{ss}}$	$\frac{L_0}{\omega_s R_s}$

equations:

$$s \cdot i_{qr} = -\left(\frac{1}{T_1}\right) \cdot i_{qr} + \left(\frac{X_2}{T_1}\right) \cdot V_{qr} \tag{6}$$

$$s \cdot \omega = -\left(\frac{X_3}{2H_t}\right) \cdot i_{qr} + \left(\frac{X_2}{2H_t}\right) \cdot T_m \tag{7}$$

$$P_e = \omega \cdot X_3 \cdot i_{qr} \tag{8}$$

For linearization, Eq. (14) can be represented as shown in the following equation:

$$P_e = \omega_{opt} \cdot X_3 \cdot i_{qr} \tag{9}$$

where ω_{opt} denotes the operating point of the rotational speed. While T_e denotes electromagnetic torque, T_m denotes the mechanical torque, ω denotes the rotational speed. Moreover, P_e denotes the active power of wind turbine, i_{qr} denotes q-axis rotor current, v_{qr} denotes q-axis rotor voltage. H_t denotes equivalent inertia of WT. Table 1 displays the applied parameters with respect to Fig. 2.

where
$L_{ss} = L_s + L_m,$
$L_{rr} = L_r + L_m,$
$L_0 = L_{rr} + \frac{L_m^2}{L_{ss}}.$

ω_s represents synchronous speed, while L_m represents the magnetizing inductance. Moreover, R_r and R_s represent rotor and stator resistances, respectively. Furthermore, L_r and L_s represent rotor and stator leakage inductances. Moreover, L_{rr} and L_{ss} denote rotor and stator self-inductances, respectively.

3 H$_\infty$ Infinity Controller

The producer of the design the H$_\infty$ control scheme mainly depends on the static or dynamic feedback controller. So, H$_\infty$ synthesis has been approved as the following:

i. Formulation: determining the weighting functions of the input and output to accomplish the robustness and performance necessities.
ii. Solution: the weighting functions may be modified to reach the optimal design of the controller.

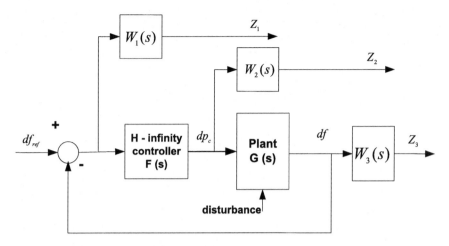

Fig. 3 Scheme of the plant with H_∞

In the proposed wind generation system including H_∞ controller, The system G(s) is the plant with weighting functions of $W_1(s)$ corresponding the signals of error while the $W_2(s)$ corresponding the signals of control and $W_3(s)$ corresponding the signals of output. The proper selecting of the weighting functions is the core of H_∞ scheme.

Considering the control system of Fig. 3. The sets of weighting functions have been chosen to ensure robust performance goals as the following.

The $W_1(s)$ assists to realize a good tracking performance, which means in this case that frequency deviation equals zero. Moreover, the other benefit of the proper choice of this weighting function is the rejecting of the disturbances. This can be expressed mathematically as

$$Z_1 = W_1(s)[0 - \Delta f] \tag{10}$$

The other weight transfer function $W_2(s)$ helps to provide the robustness to plant additive perturbations. This can be expressed mathematically as:

$$Z_2 = W_2(s) \cdot u(s) \tag{11}$$

where $u(s)$ denotes output control signals for H_∞ regulator.

Finally, the function $W_3(s)$ is essential to determine the closed-loop bandwidth and sensor noise attenuation at high frequencies, mathematically can be described as shown in the following equation:

$$Z_3 = W_3(s) \ [P] \tag{12}$$

4 Linear Quadratic Gaussian (LQG)

The LFC for a single-area power system has been established using H∞ and LQG techniques in this Chapter. The LQG is planned using the linear model of the plant, an integral objective function, and also Gaussian white noise model of disturbance and noise. Moreover, the LQG involves an optimal state feedback gain "k" and the Kalman estimator. Furthermore, the optimal feedback gain has been estimated to achieve the feedback control law u = −kx resulting in the minimization of the following performance index:

$$H = \int_0^\infty \left(X^T QX + u^T Ru\right)dt \tag{13}$$

where Q and R denote the positive definite or semidefinite Hermitian or real symmetric [15]. Moreover, the optimal state feedback u = −kx could be implemented considering the plant's full state. So, the states have been preferred to be frequency variation Δf, mechanical power change ΔP_{mi}, and governor output change ΔP_g. The Δf and ΔP_c have been measured and have been considered as inputs to the Kalman estimator. The Kalman filter has been utilized in order to observe the following unmeasured states:

$$\hat{x} = \left[\, \Delta \hat{f} \;\; \Delta \hat{p}_m \;\; \Delta \hat{p}_g \,\right]$$

The states observation has been estimated based on the following equation:

$$\left(\dot{\hat{x}}\right) = (A - Bk - LC)\hat{x} + Ly \tag{14}$$

where L denotes the Kalman gain. Moreover, the value of the L can be calculated considering the matrices of Q_n and R_n.

5 System Configuration

The control scheme of the complete system has been illustrated in Fig. 4. The system contains a simplified frequency response model for a single-area power system considering the integration of the wind energy system and including the proposed controller H∞ and LQG estimator has been illustrated in Fig. 4.

The measured and reference frequency deviation Δf and $\Delta f_{ref} = 0\,Hz$ have been applied as the inputs of H∞ controller to get the supplementary control action ΔP_c. Moreover, the frequency deviation Δf and supplementary control action ΔP_c are used as the input of the Kalman filter for estimating the states of

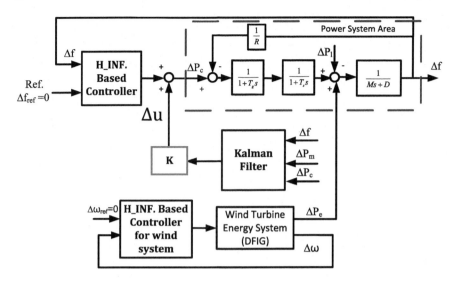

Fig. 4 Single area power system with wind energy system considering the proposed H∞ and LQG controllers

$\hat{x} = \begin{bmatrix} \Delta \hat{f} & \Delta \hat{p}_m & \Delta \hat{p}_g \end{bmatrix}$. The observed states are growing with the optimal state feedback gain "k" for getting the optimal control signal.

The subsequent set of weighting functions have been selected after many iterations in order to achieve the favorite robustness.

For the power system plant, the weighting functions are
$W_1 = \frac{s+0.3}{s+0.56}$, $W_2 = \frac{s}{s+0.04}$, and $W_3 = \frac{1}{s+2}$.
While for the wind turbine energy system are
$W_1 = 1$, $W_2 = 0$, and $W_3 = \frac{1}{s}$.

Each model of the plant and the Wind turbine DFIG has been modeled in Simulink as input–output system. Then, the MATLAB tool of *linmod* has been used to extract a continuous-time linear state-space model. This model is used to plan the robust controllers of H∞ and LQG.

6   Results and Discussions

Simulation tests are executed to confirm the effectiveness of the presented control system considering H∞ and LQG controllers in MATLAB/Simulink platform. All parameters of the system are detailed in Table 2 [2]. Moreover, the data of the DFIG WT are detailed in Table 3.

The WT involves 200 of 2 MW of VSWTs. Moreover, the parameters of WT have been specified in Table 3.

X_m denotes magnetizing reactance. Also, X_{ls} and X_{lr} denote leakage reactance of the rotor and stator, respectively.

Table 2 Single area power system parameters

D (pu/Hz)	2H (pu.sec)	R (Hz/pu)	T_g (s)	T_t (s)
0.015	0.1667	3.00	0.08	0.40

$(p_e)_{Base} = 800$ MVA

Table 3 The WT parameters [14]

Operating point (MW)	Wind speed (m/s)	Rotational speed
247	11	1.17
R_r (pu)	R_s (pu)	X_{lr} (pu)
0.00552	0.00491	0.1
X_{lr} (pu)	X_m (pu)	H_t (pu)
0.09273	3.9654	4.5

The generation rate constraint (GRC) of 10% per minute is considered. Moreover, the maximum value of the dead band for governor has been considered as 0.05 pu [2].

Case study 1
In this case of study, wind turbine participation is simulated at nominal parameters and specifications. Two tests have been carried out, the first is to test the system performance with WT. While the second test has been considering the absence of the WT. The two cases of tests have been compared with respect to the frequency deviation and the variation of mechanical power. A step load change is applied with the value of 0.02 pu at the time of $t = 50$ s. Figure 5 shows the simulation results of this case study. In this case, the parameter and data of the power plant are assumed as the nominal data. So, it is clear the performance of both two cases is stable. However, it is exposed to the results that the system is slow compared to those without the participation of WT because of the absence of the WT inertia. The assumed variation of the wind turbine speed has also been illustrated in the figure.

Case study 2
To validate the proposed H$_\infty$ + LQG controller with wind turbine participation, the dynamic performance of the control system has been simulated at nominal parameters under step load change of 0.06 pu. The simulation results of this case under study have been revealed in Fig. 6. This figure displays that the performance of the proposed controller system with wind turbine participation is better than those without wind turbine participation. The overshoot is decreasing from 0.05626 to 0.0205 with wind turbine participation. So, integration of the WT led to the enhancement of the features of the planned controller of H$_\infty$ + LQG in this case of study.

Case study 3
The system parameters are varied in this case of study for authenticating the robustness of controller performance. The system is tested under parameter uncertainty

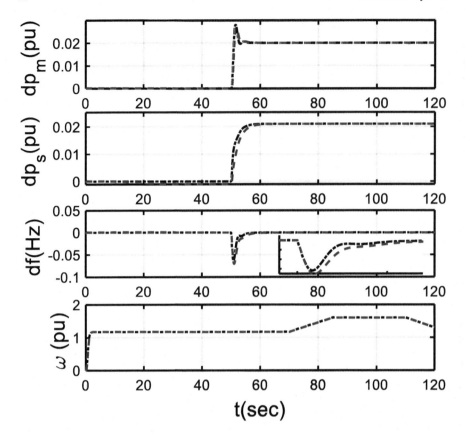

Fig. 5 Control system response for case 1; with H_∞ + LQG controller with the participation of wind turbines (dash blue curve), H_∞ + LQG controller without the participation of wind turbines (dash-dot black curve)

of the damping and inertia. The simulation results have been shown in Fig. 7. The system performance under parameter uncertainty is still stable and the proposed controller with the wind turbine participation is capable of damping the oscillations. Moreover, the features of the designed H_∞ + LQR is associated with those of the H_∞ controller only. The comparison shows that the proposed controller of H_∞ + LQR has the best performance considering overshoot.

Case study 4
In this case of study, the performance of the system has been investigated considering connecting the WT and disconnecting the WT at 40 and 100 s, respectively. The load will be assumed o be 0.01 pu at 5 s and is not changed to the end of the simulation time. The results of this case under study have been exposed in Fig. 8. The figure displays the performance of the H_∞ + LQG and conventional I controllers for the purpose of comparison and validation. The figure illustrates ΔP_m, ΔP_s, Δf, and ω from top to bottom. From the figure, the performance of the H_∞ + LQG is the best

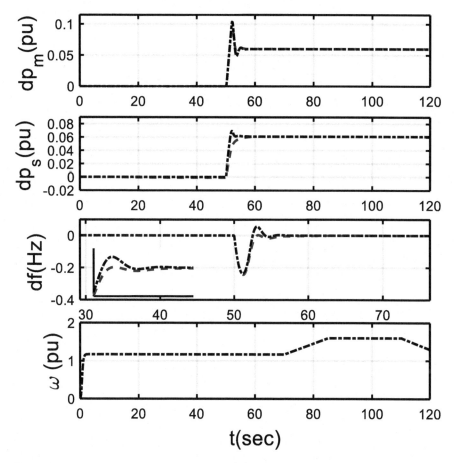

Fig. 6 Control system response for case 2; with **H∞+ LQG** controller with participation of wind turbines (dash blue curve), **H∞+ LQG** controller without participation of wind turbines (dash-dot black curve)

one concerning the fast response and lower overshooting. While the H_∞ + LQG controller has a very high performance, the conventional I controller performance is the lowest one.

7 Conclusion

In this chapter, an LFC control scheme considering the H_∞ and LQG is planned for a single-area power system bearing in mind the wind turbine energy system. The future controllers are planned to work in parallel to guarantee overall stability. Moreover, the simulation tests have been applied to validate the control system under different

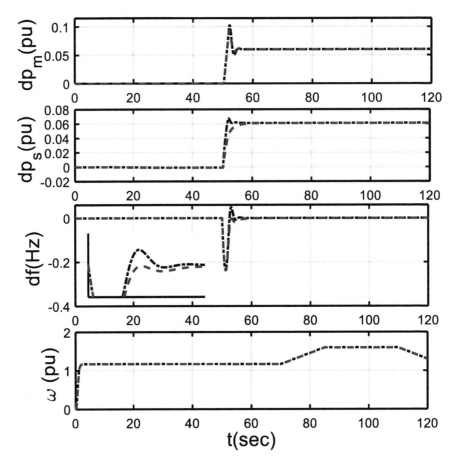

Fig. 7 Control system response for case 3; with **H∞ + LQG** controller with the participation of wind turbines (dash blue curve), **H∞** controller with the participation of wind turbines (dash-dot black curve)

load changes and parameter uncertainties. The overall performance of the proposed controller of H∞ + LQG has been compared with those of H∞ and conventional I controller. The obtained results confirmed that the planned H∞ + LQG has acceptable performance and stability. Furthermore, the designed H∞ + LQG has been proven as a robust controller in contrast to the load variation and parameter uncertainties. Also, the results specified that the combination H∞ + LQG and H∞ controllers can give the required robustness. However, the H∞ + LQG can give a better dynamic response especially in the case of considering the wind turbine energy system. Moreover, the priority and effectivity of the H∞ + LQG controller are validated in the case of severing parameter uncertainties more than H∞ controller.

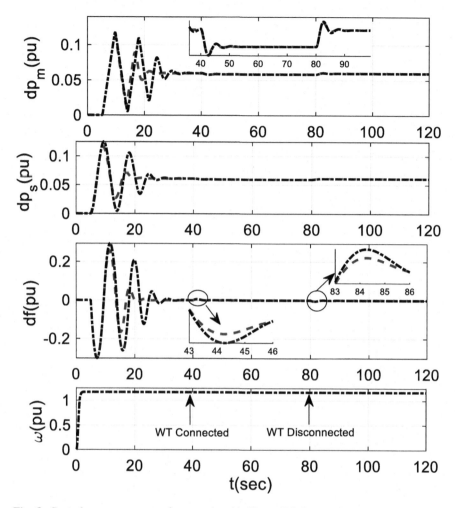

Fig. 8 Control system response for case 4; with **H∞ + LQG** controller (dash blue curve), Conventional **I** controller (dash-dot black curve)

References

1. Kundur P (1994) Power system stability and control. McGraw-Hill, New York
2. Bevrani H (2009) Robust power system control. Springer, New York
3. Eltamaly AM, Mamdooh A-S, Sayed K, Abo-Khalil AG (2020) Sensorless active and reactive control for DFIG wind turbines using opposition-based learning technique. Sustainability 12(9):3583
4. Mohamed, TH, Zaki Diab AA, Hussein MM (2015) Application of linear quadratic gaussian and coefficient diagram techniques to distributed load frequency control of power systems. Appl Sci 5(4):1603–1615
5. https://www.irena.org/. Accessed 18 March 2020

6. Holdsworth L, Ekanayake JB, Jenkins N (2004) Power system frequency response from fixed speed and doubly fed induction generator-based wind turbines. Wind Energy 7:21–35
7. Eltamaly AM, Al-Saud MS, Abo-Khalil AG (2020) Dynamic control of a DFIG wind power generation system to mitigate unbalanced grid voltage. IEEE Access 8:39091–39103
8. Abo-Khalil AG, Alghamdi AS, Eltamaly AM, Al-Saud MS, RP P, Sayed K, Bindu GR, Tlili I (2019) Design of state feedback current controller for fast synchronization of DFIG in wind power generation systems. Energies 12:2427
9. Mohamed, MA, Zaki Diab AA, Rezk H (2019) A novel adaptive model predictive controller for load frequency control of power systems integrated with DFIG wind turbines. Neural Comput Appl 1–11
10. Mullane A, O'Malley M (2005) The inertial response of induction-machine-based wind turbines. IEEE Trans Power Syst 20(3):1496–1503
11. Lee HJ, Park JB, HoonJoo (2006) Robust LFC for uncertain nonlinear power systems: a fuzzy logic approach. Inf Sci 176:3520–3537
12. Rerkpreedapong D, Hasanovic A, Feliachi A (2003) Robust load frequency control using genetic algorithms and linear matrix inequalities. IEEE Trans Power Syst 18(2):855–861
13. Demiroren A, Zeynelgil HL, Semgor NS (2001) The application of an technique to load-frequency control for three-area power system. In: Paper accepted for presentation at PPT 2001, 2001 IEEE porto power tech conference 10th–13th September, Porto, Portugal
14. Liu F, Song YH, Ma J, Mai S, Lu Q (2003) Optimal load-frequency control in restructured power systems. IEE Proc Gener Trans Distrib 150(1):87–95
15. Wang Y, Hill DJ, Guo G (1998) Robust decentralized control for multimachine power system. IEEE Trans Circuits Syst Fundam Theory Appl 45(3)
16. Stankovic AM, Tadmor G, Sakharuk TA (1998) On robust control analysis and design for load frequency regulation. IEEE Trans Power Syst 13(2):449–455
17. Pan CT, Liaw CM (1989) An adaptive controller for power system load-frequency control. IEEE Trans Power Syst 4(1):122–128
18. Cam E, Kocaarslan I (2005) Load frequency control in two area power systems using fuzzy logic controller. Energy Convers Manag 46:233–243
19. Birch AP, Sapeluk AT, Ozveren CS (2005) An enhanced neural network load frequency control technique. Control 94:21–24, Conference Publication No. 389, IEE 1994. ASCE J 5(II)
20. Abdel-Magid YL, Dawoud MM (1995) Genetic algorithms applications in load frequency control. In: Genetic algorithms in engineering systems: innovations and applications, 12–14 September 1995, conference publications No. 414, IEE
21. Diab AAZ, Al-Sayed A-HM, Mohammed HHA, Mohammed YS (2020) Development of adaptive speed observers for induction machine system stabilization. Springer
22. Abo-Khalil AG, Alghamdi A, Tlili I, Eltamaly AM (2019) Current controller design for DFIG-based wind turbines using state feedback control. IET Renew Power Gener 13(11):1938–1948
23. Eltamaly AM, Mohamed YS, El-Sayed A-HM, Elghaffar ANA (2019) Analyzing of wind distributed generation configuration in active distribution network. In: 2019 8th international conference on modeling simulation and applied optimization (ICMSAO), pp 1–5 (2019)
24. Eltamaly AM, Mohamed YS, El-Sayed A-HM, Elghaffar ANA (2019) Reliability/security of distribution system network under supporting by distributed generation. Insight-Energy Sci 2(1):1–14
25. Eltamaly AM, Farh HM (2013) Maximum power extraction from wind energy system based on fuzzy logic control. Electr Power Syst Res 97:144–150
26. Eltamaly AM, Alolah AI Abdel-Rahman MH (2011) Improved simulation strategy for DFIG in wind energy applications. Int Rev Model Simul 4(2)
27. Eltamaly AM (2007) Modeling of wind turbine driving permanent magnet generator with maximum power point tracking system. J King Saud Univ-Eng Sci 19(2):223–236
28. Eltamaly AM, Alolah AI, Abdel-Rahman MH (2010) Modified DFIG control strategy for wind energy applications. In: SPEEDAM, pp 653–658. IEEE
29. Eltamaly AM, Alolah AI, Farh HM, Arman H (2013) Maximum power extraction from utility-interfaced wind turbines. New Dev Renew Energy 159–192

D-STATCOM for Distribution Network Compensation Linked with Wind Generation

Ali M. Eltamaly, Yehia Sayed Mohamed, Abou-Hashema M. El-Sayed, Amer Nasr A. Elghaffar, and Ahmed G. Abo-Khalil

Abstract Wind generation is considered as one of the optimum renewable energy sources due to its more saving running cost, zero-emission, and friendly environment at comparing with the traditional power plants. Using the wind farms with the utility grid can't reach the optimum compensation during any abnormal condition in the power system. Distributed Static Synchronous Compensators (D-STATCOM) is a power electronic control system to be used with the distribution power network for harmonics current elimination, reactive power compensation, voltage regulation, voltage flicker mitigation, and frequency regulation. D-STATCOM provides effective compensation to the unbalance or the nonlinear loads by injecting the required accurate value at the Point of Common Coupling (PCC) depending on the Voltage Source Converter (VSC). This chapter proposes a design procedure of a high-power D-STATCOM with the distribution network linked with the wind generation for enhancing the power system quality. The proposed simulation in this chapter has been done using the MATLAB/Simulink software for distribution voltage control, loadability, and power loss reduction using the D-STATCOM control techniques with the electrical distribution network.

Keywords D-STATCOM · Voltage compensation · Distribution power networks · Power quality

A. M. Eltamaly (✉)
Electrical Engineering Department, Mansoura University, Mansoura, Egypt
e-mail: eltamaly@ksu.edu.sa

Sustainable Energy Technologies Center, King Saud University, Riyadh 11421, Saudi Arabia

K.A. CARE Energy Research and Innovation Center, Riyadh 11451, Saudi Arabia

Y. S. Mohamed · A.-H. M. El-Sayed · A. N. A. Elghaffar
Electrical Engineering Department, Minia University, Minia, Egypt

A. G. Abo-Khalil
Department of Electrical Engineering, College of Engineering, Majmaah University, Almajmaah 11952, Saudi Arabia

Department of Electrical Engineering, College of Engineering, Assiut University, Assiut 71515, Egypt

© The Author(s), under exclusive license to Springer Nature Switzerland AG 2021
A. M. Eltamaly et al. (eds.), *Control and Operation of Grid-Connected Wind Energy Systems*, Green Energy and Technology,
https://doi.org/10.1007/978-3-030-64336-2_5

Nomenclature

R	Rotor resistance impedance
U	Output voltage
X1	Stator reactance
X2	Rotor reactance
X_σ	Total reactance rotor and stator
S	Generator slip
Q	Reactive Power
Q_C	The parallel capacitor group
Q_{N-Unit}	Reactive power capacitor compensation capacity
[n]	Assuming actual capacitor investment group
Cp max	Optimal power coefficient
λ	Ratio between the wind speed and the rotor speed
v_w	Wind speed
w	Wind turbine rotational speed
P_m	The extracted mechanical power by the constant pitch
ρ	Air density
C_p	Turbine coefficient
A_r	Area swept by the blades
w_t	The rotational speed
k_{opt}	The unique parameter
L_f	Interfacing inductors
m	The modulation index
w	The system frequency
vA1	Output injection voltage inverter 1
vA2	Output injection voltage inverter 2
i_{A1}	Currents delivered by inverter-1
i_{A2}	Currents delivered by inverter-2
L_0	Mutual inductance
L_{lk}	Leakage inductance
e_{gr}	Equivalent grid voltage
L_{gr}	Equivalent inductance
R_{gr}	Equivalent resistance
v_{gr}	Voltage at the point of common coupling (PCC)
i_{gr}	The current at PCC
v_{A1}	The instantaneous voltage generates by converter-1
v_{A2}	The instantaneous voltage generates by converter-2
v_f	The voltage over filter capacitor c_f

1 Introduction

Doubtless the electrical power system network has a lot of challenges to save the stable system especially with the extension and the increasing of the nonlinear loads. Using the compensation devices for enhancing the system to reach the optimum operation is the solution to save the electrical system in stable condition [1]. The power system quality is affected by the disturbances or any abnormal condition as voltage sag/swell, which directly have an impact on electronic devices or any critical loads causing heating or malfunction of devices and the high cost related to the system shut down. With the extension of the demands of the wind energy to utilize the natural environment and to save the generation running costs, the reactive/voltage control in the distribution system is becoming more important [2]. Wind turbines utilizing the frequency converters are generally fit for controlling the reactive power to zero or potentially of providing or devouring reactive power as indicated by needs, even though this is restricted by the size of the converter [3]. So, it's important to use compatible devices for enhancing the system quality during abnormal conditions [4, 5]. However, unfortunately of using the compensation techniques to finding other bad affection as feeding harmonics by the compensation devices or noncontinuous compensation or finding the inrush current [6]. The system compensation by a fixed shunt capacitor is a simple way, but it's not accurate especially with the fast load variation and with the critical loads. Following the improvement of the power electronics circuits and the advanced methodology, the Flexible AC Transmission Systems (FACTS) devices can solve the power system quality depending on the thyristors technologies [7]. FACTS devices are consisting of the optimum way to operates in fast, accurate compensation techniques, and easy to install with the system for control the voltage profile, damp the power system oscillation [6–8]. Consequently, Custom Power Devices (CPD) are preferred for the electrical power quality improvement due to the lower cost, greater flexibility, and increase of system security [9]. Static Synchronous Compensation (STATCOM) control techniques and Static Var Compensation (SVC) modules are the famous compensation ways of the FACTS families [10–15]. Distributed Static Synchronous Compensators (D-STATCOM) is a shunt compensator device to be installed with the electrical distribution network for voltage control, also it can tackle the power quality issues [16–24]. The D-STATCOM configuration is adaptable enough to abuse multi-usefulness coupling with the power system. Besides, D-STATCOM is a multi-functional gadget that simple to give viable remuneration to the non-direct loads and unbalance stacks by infusing suitable responsive power contingent upon the thyristors for exchanging [21]. Figure 1 shows the streamlined structure to connect the D-STATCOM with the distribution network for improving the PQ. There are many designs of the D-STATCOM multilevel inverter control as Cascaded H-Bridge [22] and Diode-Clamped [23]. However, using a new control technique with the D-STATCOM topology by adding one more converter with the D-STATCOM to increase the voltage source converter power. This chapter discusses the importance of using the D-STATCOM with the distribution system for voltage control depending

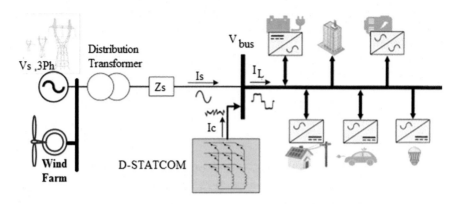

Fig. 1 D-STATCOM for PQ control

on the Voltage Source Converter (VSC) for the optimum power quality improvement. The proposed analysis in this chapter has been validated with the distribution network to compensate for the voltage on a 25-kV distribution network using the MATLAB/SIMULINK software to shows the voltage compensation during the fault condition in the power system.

2 Wind Generation Output Power

Utilizing the environment renewable sources like the wind can be used for rotating the generator to find free electric power without running cost as comparing with the fossil fuel generation system. By the last simple discussion, Fig. 2 shows the fundamental wind turbine parts which depending upon the transformer to connect with the utility grid [24–30]. Moreover, the typical technique to change over the low- and high-speed conditions and to generate the electric power from mechanical power a gearbox and a generator is required to be synchronized with electric utility network. The gearbox changes the low speed of the turbine rotor to high speed. Notwithstanding some types

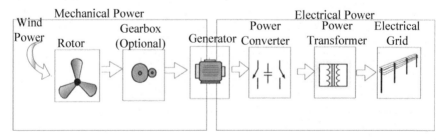

Fig. 2 Main component for WG

Fig. 3 The equivalent simplification model for WG

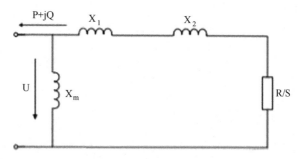

of wind turbines like the one use permanent magnet and multipole generator systems are not using gearbox [31]. The wind energy system is considered as precarious for power due to its compelled consistency and irregularity [32]. By using the exchange plans with nearby generating power plants, the power system can disentangle the fluctuations caused by the wind energy systems. At the point when the generated power from the wind is lower than the loads, the primary game plan is used to use standard generation units to cover this shortage of wind power [33]. The wind generation is relying upon the criticism pay through the utility grid, thus, the wind generation can be classified as PQ or PV busbar types in the power stream modeling [34–38]. The coordinated generator circuit can be streamlined to the proportionate circuit as appeared in Fig. 3. Where, the stator resistance can be disregarded, R is the rotor resistance impedance, U is the yield voltage, X1, X2 are the stator, and rotor reactance impedance individually and Xm is the excitation reactance.

Equation (1) shows the total reactance rotor and stator, to uses for calculating the active generated power in Eq. (2).

$$X_\sigma = X_1 + X_2 \tag{1}$$

$$P = \frac{SRU^2}{S^2 X_\sigma^2 + R^2} \tag{2}$$

With consideration, S is the generator slip, that is related to Eq. (3) to obtain the reactive power by using Eq. (4).

$$S = R\left(U^2 - \sqrt{U^4 - 4X_\sigma^2 P^2}\right)/2PX_\sigma^2 \tag{3}$$

$$Q = \left[R^2 + X_\sigma(X_m + X_\sigma)S^2\right]/SRX_m \tag{4}$$

There are various methods when choosing the reactive power compensation system inside the WG gatherer design, the parallel shunt capacitors with the WG can compensate for the reactive power [39–42]. Which able to calculates by the below equation number (5) to find the total required capacitor group value Q_C from Eq. (6).

$$Cos(\varnothing) = P/\sqrt{P^2(Q_C - Q)^2} \qquad (5)$$

$$Q = P\left[\sqrt{\frac{1}{(Cos(\varnothing 1))^2} - 1} - \sqrt{\frac{1}{(Cos(\varnothing 2))^2} - 1}\right] \qquad (6)$$

Chosen the compatible and prober way should follow the prerequisites of the WG plant. Further things for thought may incorporate voltage impediments, especially during the switching of shunt capacitors, power quality necessities, for example, glint during fire up or cut-in, harmonics, and so forth [43, 44]. By expecting the real capacitor speculation bunch is [n], and the receptive power capacitor pay limit is Q_{-}(N-Unit), which mimics at the evaluated voltage, which can use to generate the WG reactive power from Eq. (8).

$$[n] = Q_C/Q_{N-Unit} \qquad (7)$$

$$Q' = Q_C - Q \qquad (8)$$

The controller used for WG speed control essential function is the Maximum Power Point Tracking (MPPT) of the available power in the wind. An expanding number of bigger WGs (starting from 1 MW and more) are created with a functioning slow down power control component. Actually, the dynamic slow down turbines take after pitch-controlled turbines since they have pitch capable blades. To get a sensibly huge force (turning power) at low wind speeds, the WG will ordinarily be customized to pitch their blades a lot of like a pitch-controlled wind turbine at low wind speeds. Regularly they utilize just a couple of fixed advances depending on the wind speed. At the point when the turbine arrives at its evaluated power, nonetheless, it will see a significant distinction from the pitch-controlled wind turbines: If the generator is going to be over-burden, the turbine will contribute its blades the other way from what a pitch-controlled WG does. As such, it will build the approach of the rotor blades to cause the blades to go into a more profound slow down, in this manner squandering the overabundance energy in the wind [45]. One of the active stalls of a functioning slowdown is that one can control the active power more precisely than with passive stall, to abstain from overshooting the appraised power of the turbine toward the start of a whirlwind. Another preferred position is that the WG can be run precisely at the evaluated power of the machine at all high wind speeds [43–45]. Figure 4 shows the variable wind speed [43]. Where, (λ) is the tip-speed to simulate the ratio between the wind speed and the rotor speed as shown in Eq. (9).

$$\lambda = \frac{R_p w}{v_m} \qquad (9)$$

with considering, is the wind speed represented by v_w and the WG speed is w [46]. Hence, the extracted mechanical power by the constant pitch can be calculated from

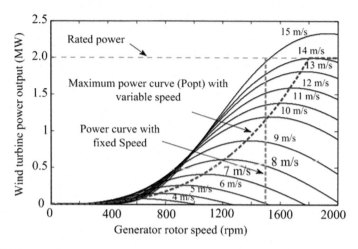

Fig. 4 Wind generation output power during fixed and variable wind speed

the below Eq. (10):

$$P_m = \frac{1}{2} \cdot \rho A_r v_w^3 \cdot C_p(\lambda) \tag{10}$$

where; ρ is the air density is presented by ρ, the turbine coefficient presented by C_p and the area swept by the blades can simulate by A_r. The wind velocity V1, V2 are presented for WG control curve as shown in Fig. 5. With a note, at the WG

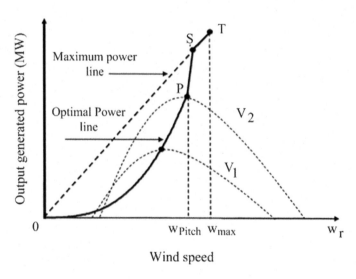

Fig. 5 WG power control curve

operates at optimal power coefficient (Cp max) in this simulation, at each velocity, there is one maximum power capture point [47]. Moreover, an individual optimal power available for the WG speed by using the below Eq. (11).

$$P_{opt} = k_{opt} w_T^3 \tag{11}$$

where, w_t is the rotational speed and k_{opt} is the unique parameter.

3 D-STATCOM Control Strategies

D-STATCOM control techniques are designed to link with the three-phase power system by three-phase stacks as shown in Fig. 1. The D-STATCOM is a shunt connected custom power device depending on the Direct Current (DC) energy storage for injection of the voltage through the coupling transformer under control by the VSC converts and the output filter. The point of adding the VSC to convert the DC stockpiling batteries voltage to the decent three-phase AC yield voltages. The created voltages are in-phase and interconnected with the utility grid through a coupling transformer. D-STATCOM legitimate yield setting of the phase angle and the voltage greatness permits successful control of active and reactive power flow between the D-STATCOM and the utility distribution system [48]. The three-phase loads may be of various sorts like lagging power factor, unbalance, linear or nonlinear loads, and joined among linear and nonlinear loads. However, to reduce the swiping in the compensating current, it requires to use the interfacing inductors (L_f) at the AC side of the VSC. The operation characteristics of the D-STATCOM can be justified depending on the dc-link capacitor, equivalent transformer. An insulated-gate bipolar transistor (IGBT)-based VSC, and filter resistance, and filter inductance, and the utility grid as a three-phase source. Figure 6 shows the equivalent D-STATCOM circuit, which can use to find the combination of the PCC voltage and the inverter-generated voltage for system compensation.

The simplified equation for phase (A), phase (B), and phase (C) can find from Eqs. (12), (13), (14), respectively.

$$R_s i_a + L_s \left[\frac{di_a}{dt} \right] = V_{Pa} - V_{ca} \tag{12}$$

$$R_s i_b + L_s \left[\frac{di_b}{dt} \right] = V_{Pb} - V_{cb} \tag{13}$$

Fig. 6 D-STATCOM equivalent simplified circuit

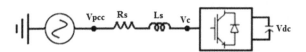

$$R_s i_c + L_s \left[\frac{di_c}{dt} \right] = V_{Pc} - V_{cc} \tag{14}$$

By transforming the equations to obtain the below equations, with considering m is the modulation index, the frequency is w.

$$L_s \left[\frac{di_d}{dt} \right] + R_s i_d = V_{Pd} - m V_{dc} \cos\theta + L_s w_{iq} \tag{15}$$

$$L_s \left[\frac{di_q}{dt} \right] + R_s i_{dq} = V_{Pq} - m V_{dc} \sin\theta - L_s w_{id} \tag{16}$$

where from Eqs. (15) and (16) can be justified as shown in Eq. (17).

$$\frac{d}{dt} \begin{bmatrix} id \\ iq \end{bmatrix} = \begin{pmatrix} -R_s/L_s & w \\ -w & -R_s/L_s \end{pmatrix} \begin{bmatrix} id \\ iq \end{bmatrix} + \frac{1}{L_s} \begin{pmatrix} V_{pd} - V_{cd} \\ V_{pq} + V_{cq} \end{pmatrix} \tag{17}$$

The simplified equations are simplified by ignoring the voltage harmonics as shown in the following.

$$m.V_{dc} \cos\theta = V_{cd} \tag{18}$$

$$m.V_{dc} \sin\theta = V_{cq} \tag{19}$$

By considering the inverter is a lossless circuit, so the instantaneous power can be found by the power balance theory at the ac-dc terminal to find as shown in Eq. (20) and the dc side circuit can find as shown in Eq. (21).

$$V_{dc}.i_{dc} = \frac{3}{2}(V_{cd}.id + V_{cq}.iq) \tag{20}$$

$$i_{dc} = C.\frac{d(V_{cd})}{dt} = \frac{3}{2}.m.(id.\cos\theta - iq.\sin\theta) \tag{21}$$

By combining Eq. (17) and Eq. (21)

$$\frac{d}{dt} \begin{pmatrix} i_d \\ i_q \\ V_{dc} \end{pmatrix} = F \begin{pmatrix} i_d \\ i_q \\ V_{dc} \end{pmatrix} - \frac{1}{Ls} \begin{pmatrix} i_d \\ i_q \\ V_{dc} \end{pmatrix}$$

where, F is given as:

$$\begin{pmatrix} -R_s/L_s & w & \frac{-m}{L_s}\cos\theta \\ -w & -R_s/L_s & \frac{m}{L_s}\sin\theta \\ \frac{3}{2}\frac{m}{c}\cos\theta & \frac{-3}{2}\frac{m}{c}\sin\theta & 0 \end{pmatrix} \tag{22}$$

At this point, to illustrate the active and reactive power for D-STATCOM compensation, it can simplify by the below Eqs. (23), (24), respectively.

$$P = V_{pd.id} + V_{pd.iq} = V_{pd.id} = V_{pd.id} \tag{23}$$

$$q = V_{pq.id} - V_{pq.iq} = -V_{pd.iq} = -V_{pd.iq} \tag{24}$$

Finally, the D-STATCOM module operates to compensate the system depending on controlling of *id*, *iq* as shown in Eq. (23) for active power injection and Eq. (24) for the reactive power injection to the distribution busbar.

4   D-STATCOM Characteristic Logic

Depending on the coupling of magnetic elements for the parallelization of standard two-level or three-level inverters using the D-STATCOM dual-converter configuration can be created to increase the VSC output power as shown in Fig. 7 for the dual-converter with common DC link and Fig. 8 for the dual-converter with isolated DC link [49, 50]. According to the coupling between the two converters, an equal

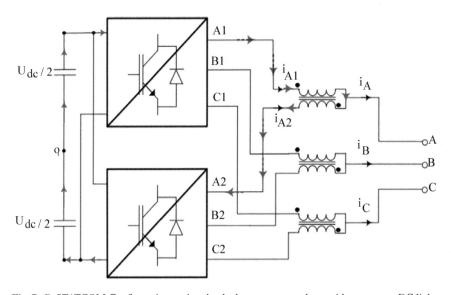

Fig. 7 D-STATCOM Configurations using the dual-converter topology with a common DC link

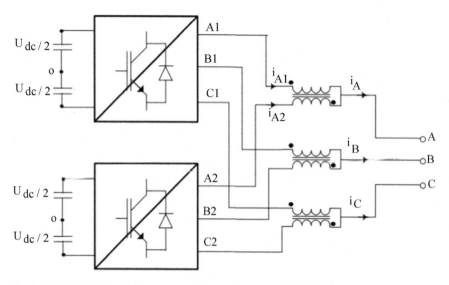

Fig. 8 D-STATCOM the dual-converter design with separated DC link

current will flow from both converters, which can be known as a parallel interleaved VSC [51]. The harmonic injection can be significantly reduced by the interleaved carrier signal of both SVCs, which is phase-shifted by 180° [51–53]. With notes, there is a potential difference between the output injection voltage V_{A1} and V_{A2} due to carriers are interleaved.

The circulating current is depressed by the effection of the DC links which is described in Figs. 7 and 8. Where VSC contains two parallel parts of three-phase converters. The parallelization of inverter phases is accomplished by the methods for two neighboring coils that are attractively connected. Ignoring copper losses, the mutual induction between the coupled coils can be portrayed by the equivalent circuit for Phase-A in Fig. 9, where the currents delivered by each inverter is i_{A1} and i_{A2}, to express as a function of common-mode current i_{mc} and differential-mode current i_{md} (circulating current) as shown in Eqs. (25) and (26).

$$i_{A1} = \frac{i_{mc}}{2} + i_{md} \tag{25}$$

$$i_{A2} = \frac{i_{mc}}{2} - i_{md} \tag{26}$$

With a note, the two inductors in the same phase are the same number of turns and arranged on the same ferromagnetic core [51]. The mutual inductance L_0 and the leakage inductance L_{lk} can be expressed in function of the self-inductance, where k expresses the coupling factor between the coils that can equalize "1" at the two coils symmetrically, to be used as shown in Eqs. (27) and (28):

Fig. 9 D-STATCOM
equivalent circuit

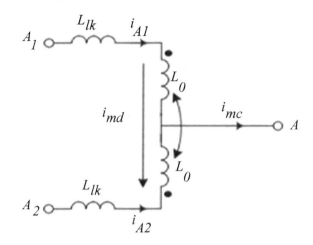

$$L_0 = kL \tag{27}$$

$$L_{lk} = (1 - k)L \tag{28}$$

The two coils are connecting to cancel the induced magnetic field as shown in the polarity points for both coils as shown in Fig. 9, to find the difference potential between the two points A1 and A2 as describes in Eq. (29).

$$v_{A1o} - v_{A2o} = (2L_{lk} + 4L_0)\frac{di_{md}}{dt} \tag{29}$$

where, the inductive reactance $L_0\omega$ seen by differential-mode currents as the quarter value of the impedance. The circulating currents can be decreased by the separation of DC links as shown in the equivalent circuit diagram in Fig. 10. DC link isolates to repress the zero-sequence track, and zero-sequence circulating current to be no longer among the two converters [51].

LCL filter is the common coupling part between the electrical system and D-STATCOM. Figure 11 shows the equivalent circuit for single-phase D-STATCOM with an LCL filter with neglecting the copper losses. Where, e_{gr} is the equivalent grid voltage, L_{gr} is the equivalent inductance, v_{gr} is the voltage at PCC point, R_{gr} is the total equivalent resistance, and i_{gr} is the current at PCC. Equations (30) to (34) shows the differential equations for the filter dynamics.

$$v_{A1} = L_{lk}\frac{di_{A1}}{dt} + L_0\frac{di_{A1}}{dt} - L_0\frac{di_{A2}}{dt} + v_A \tag{30}$$

$$v_{A2} = L_{lk}\frac{di_{A2}}{dt} + L_0\frac{di_{A2}}{dt} - L_0\frac{di_{A1}}{dt} + v_A \tag{31}$$

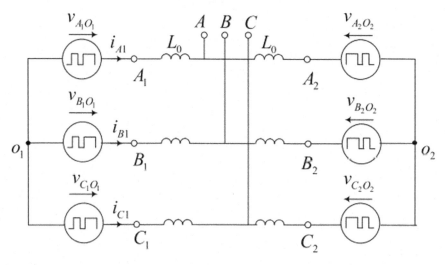

Fig. 10 Isolated DC link for D-STATCOM dual-converter

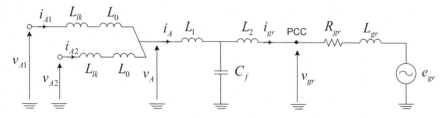

Fig. 11 Single-phase D-STATCOM equivalent circuit with LCL filter

$$v_A = L_1 \frac{di_A}{dt} + v_f \tag{32}$$

$$v_f = (L_2 + L_{gr}) \frac{di_{gr}}{dt} + R_{gr} i_{gr} + e_{gr} \tag{33}$$

$$\frac{v_{A1} + v_{A2}}{2} = \frac{L_{lk}}{2} \frac{di_A}{dt} + v_A \tag{34}$$

where, v_{A1} is the instantaneous voltage generates by converter-1, v_{A2} is the instantaneous voltage generates by converter-2, and v_f is the voltage over the filter capacitor c_f.

The common-mode voltage is defined by:

$$v_{mc} = \frac{v_{A1} + v_{A2}}{2} \tag{35}$$

The thunderous frequency is picked lower than the tweak frequency of the double converter, and the shunt capacitors of the LCL filter ought to be picked as little as conceivable because they draw receptive power, this part is fully discussed in [51].

5 D-STATCOM Simulation

D-STATCOM control techniques are used with the distribution power network to manage voltage during abnormal conditions. Figure 12 shows an example of an electrical system contains three busbars, variable loads linked at busbar-B2 and busbar-B3, two feeder lines (21 and 2 km) to fed the lads, and D-STATCOM module linked with 25 kV system. The shunt capacitor bank is used for power factor correction at busbar-B2. The variable loads at busbar-B3 are operating through a step-down transformer 25 kV/600 V to a plant retaining constantly evolving currents, in this way creating voltage flicker. At 5.0 Hz frequency, can regulate the variable magnitude of the load current which obvious the apparent power between 1.0 and 5.2 MVA, with saving the power factor at 0.9 lag. The load variation in this simulation will allow us to describe the capacity of the D-STATCOM to compensate voltage which can add capacitance value or reactance value depending on the required action as shown in Fig. 13. The D-STATCOM is connected with busbar-B3 to compensate for the reactive power by generates the in-phase voltage by a voltage source PWM inverter with the utility grid [54].

5.1 D-STATCOM Operation Characteristics

During this simulation, the load is starting a constant value to observe the D-STATCOM response to step changes in source voltage [55]. With notes, the modulation of the variable load is not in service, which depending on the Time-on and Time-off = [0.15 1] *100 > simulation stop time. The voltage source block is adjusted to simulate the internal voltage of 25.0 kV; however, the voltage is starting at 1.077pu to keep the floating injection by the D-STATCOM at busbar-3 by 1.0 pu as comparing

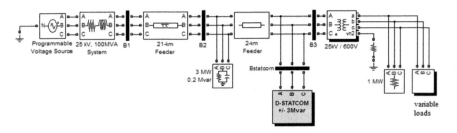

Fig. 12 D-STATCOM simulation with distribution network

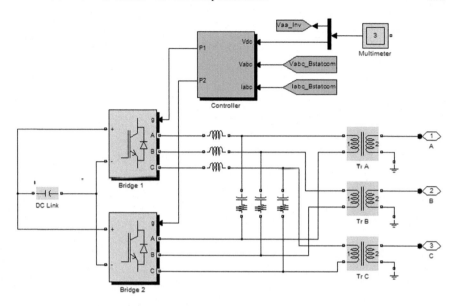

Fig. 13 D-STATCOM design with 2 inverters

with the reference value 1.0 pu. The D-STATCOM is programmed to operates by three steps at 0.2 s, 0.3 s, and 0.4 s for increasing the voltage source by 6.0% or decrease the voltage by 6.0% and return to be 1.077 pu. At 0.15second, the transient happened in the system at so the D-STATCOM will start to compensate the system at 0.2 s to increase the voltage by 6% by absorbing the system reactive power by $Q = +$ 2.7 Mvar as shown in Fig. 14. At time $= 0.3$ s the voltage source is decreased by 6.0% from the corresponding voltage to reach the reactive compensation 0.0 Mvar. Additionally, the D-STATCOM is required to compensate the system voltage to reach the reference value 1.0 p.u to change the reactive compensation from inductive to capacitive between -2.8 Mvar and 2.7 Mvar, as shown in Fig. 15 for the PWM inverter 0.56 p.u to 0.9 which corresponds to a proportional increase in inverter voltage [54, 55]. Figure 16 shows the fast reversing of reactive power at one cycle by the D-STATCOM current.

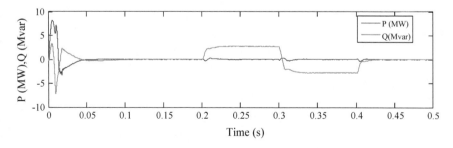

Fig. 14 D-STATCOM consumes the system reactive power ($Q = +$ 2.7 Mvar)

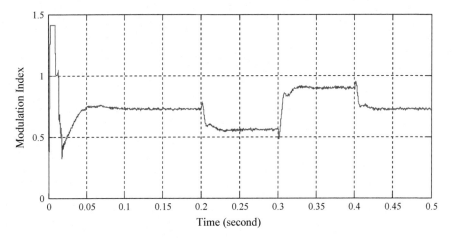

Fig. 15 D-STATCOM performed by PWM from 0.56 p.u to 0.9

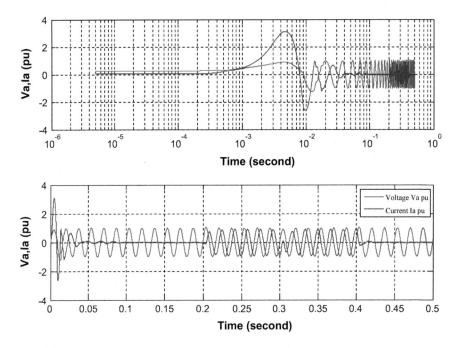

Fig. 16 Shows the fast reversing of reactive power at one cycle by the D-STATCOM current

5.2 Mitigation of Voltage Flicker

To simulate the operation characteristics of D-STATCOM for voltage compensation, it needs to variate the loads' value. By changing the "Time Variation of" In the Programmable Voltage Source block menu parameter to "None," with set the Modulation Timing Parameter in the Variable Load block menu to [Ton Toff] = [0.15 1] (without 100-multiplication factor). Also, change the "Mode of operation" parameter to "Q regulation" in the D-STATCOM control part, and set the reference reactive power Q value as zero. In this mode, D-STATCOM compensation techniques are not affecting the voltage. After running the system and check the operation for D-STATCOM as shown in Fig. 17 to verify the active power P and reactive power Q at busbar-B3, moreover, busbar-B1, and B3 voltages are simulated in Fig. 18. In another hand, without adding the D-STATCOM, the voltage at busbar-B3 will vary in range (0.96:1.04) pu [54, 55]. But, after adding the D-STATCOM control techniques, the voltage at busbar-B3 is reduced to ±0.7%. In Fig. 19, D-STATCOM compensates the voltage by injecting a reactive current which formed at 5.0 Hz as depending on

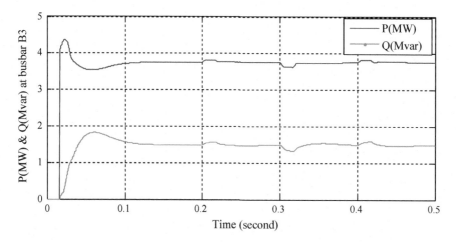

Fig. 17 For variations of P and Q at busbar-B3

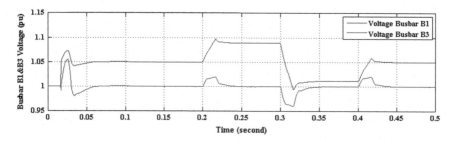

Fig. 18 Voltage value at busbar-B1 and busbar-B3

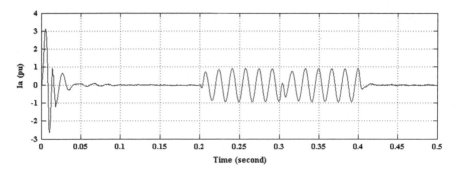

Fig. 19 D-STATCOM injection reactive current modulated at 5.0 Hz

the varying between capacitive and reactive related to the required compensation at the under/over voltage, respectively.

6 Conclusion

The blend of renewable energy sources like solar or wind energy to the electrical grid affects stability and system quality. The power system designers have to ensure a stable system and their services to retain the system in the stable region at any abnormal condition in the system. Using D-STATCOM techniques with the distribution network can achieve for enhancing the PQ. Adding the battery and the capacitor bank are used as the DC busbar in the design of D-STATCOM. In this chapter, the D-STATCOM has been demonstrated for improving power system quality during various operational events. The abnormal conditions in the distribution system as the tripping/reclosing of distribution feeder, energizing high-power transformer, switching ON/OFF of capacitor, or inductive-resistive loads. This chapter investigated one of the FACTS modules of D-STATCOM to be linked with the distribution network as a fully inductive/capacitive load as required for compensation under the support by the battery bank to be used for PQ enhancement. Finally, in this chapter, the simulation analysis was carried by MATLAB/Simulink software to discusses the operation steps of D-STATCOM to reach the optimum compensation for saving the voltage in the required stable region.

Acknowledgements The authors wish to acknowledge Alfanar Company, Saudi Arabia for supporting to complete this research, especially thanks to Mr. Amer Abdullah Alajmi (General Manager, Alfanar Engineering Service, Saudi Arabia) and Mr. Osama Morsy (Executive Manager, Alfanar Testing and Commissioning, Saudi Arabia).

References

1. Eltamaly AM et al (2020) Adaptive static synchronous compensation techniques with the transmission system for optimum voltage control. Ain Shams Eng J. https://doi.org/10.1016/j.asej.2019.06.002
2. Eltamaly AM, Elghaffar ANA et al (2018) Enhancement of power system quality using PI control technique with DVR for mitigation voltage sag. In: 2018 Twentieth international middle east power systems conference (MEPCON), Cairo, Egypt; 12/2018. https://doi.org/10.1109/MEPCON.2018.8635221
3. Eltamaly AM, Sayed Y, Abou-Hashema A, Elghaffar ANA (2020) Small-scale wind distributed generation with the power distribution network for power quality control. In: Emerging technologies and their applications in the service of sustainable development in the Arab world. 21–23 June 2020
4. Chen Z et al (2009) A review of the state of the art of power electronics for wind turbines. IEEE Trans Power Electron 24(8) (2009)
5. Ackermann T (2005) Wind power in power systems. Wiley, Book
6. Sayed Y, Abou-Hashema M, Elghaffar ANA, Eltamaly AM (2019) Multi-shunt VAR compensation SVC and STATCOM for enhance the power system quality. J Electr Eng 19
7. Brito MEC et al Adjustable VAr compensator with losses reduction in the electric System. Elec Eng J https://doi.org/10.1007/s00202-018-0692-x
8. Bindumol EK (2017) Impact of D-STATCOM on voltage stability in radial distribution system. In: International conference on energy, communication, data analytics and soft computing (ICECDS-2017)
9. Mehrdad Ahmadi Kamarposhti and Mostafa Alinezhad (2010) Comparison of SVC and STATCOM in static voltage stability margin enhancement. Int J Electr Electron Eng 4:5
10. Duarte SN et al (2020) Control algorithm for DSTATCOM to compensate consumer-generated negative and zero sequence voltage unbalance. Int J Electr Power Energy Syst 120. https://doi.org/10.1016/j.ijepes.2020.105957
11. Elghaffar ANA, Eltamaly AM, Sayed Y, Elsayed AHM (2018) Advanced control techniques for enhance the power system stability at OOS condition. Energy Sci J
12. Eltamaly AM, Mohamed YS, El-Sayed AHM, Elghaffar ANA (2019) Impact of distributed generation (DG) on the distribution system network. Ann Fac Eng Hunedoara 17(1):165–170
13. Eltamaly AM, Sayed Y, El-Sayed AHM, Elghaffar ANA (2018) Optimum power flow analysis by Newton raphson method, a case study. Ann Fac Eng Hunedoara 16(4):51–58s
14. Eltamaly AM, Sayed Y, El-Sayed AHM, Elghaffar ANA (2018) Mitigation voltage sag using dvr with power distribution networks for enhancing the power system quality. IJEEAS J 2600–7495
15. Elghaffar AA, Eltamaly A, Sayed Y, El-Sayed A-H (2018) Enhancement of power system quality using static synchronous compensation (STATCOM). EISSN, IJMEC, pp 2305–2543
16. Kamalakannan C et al (eds) Power electronics and renewable energy systems, lecture notes in electrical engineering 326. Chapter 40, Improvement of power quality in distribution system Using D-STATCOM. https://doi.org/10.1007/978-81-322-2119-7_40
17. Eltamaly AM, Sayed Y, Mustafa AH, Elghaffar ANA (2018) Multi-control module static VAR compensation techniques for enhancement of power system quality. Ann Fac Eng Hunedoara 16(3):47–51
18. Elghaffar NA, Eltamaly AM, Sayed Y, El-Sayed AHM (2018) The effectiveness of dynamic voltage restorer with the distribution networks for voltage sag compensation. Int J Smart Electr Eng 7(02):61–67
19. Eltamaly AM (2009) Harmonics reduction techniques in renewable energy interfacing converters. Renew Energy Intechweb
20. Eltamaly AM (2012) Novel third harmonic current injection technique for harmonic reduction of controlled converters. J Power Electron 12(6):925–934
21. Vural AM (2016) Self-capacitor voltage balancing method for optimally hybrid modulated cascaded H-bridge D-STATCOM. IET Power Electron 9:2731–2740

22. Gawande SP, Ramteke MR (2014) Three-level NPC inverter based new DSTATCOM topologies and their performance evaluation for load compensation. Int J Electr Power Energy Syst 61:576–584

23. Yang J et al (2013) The impact of distributed wind power generation on voltage stability in distribution systems. In: (APPEEC) conference

24. Eltamaly AM, Mohamed YS, El-Sayed AHM, Elghaffar ANA (2019) Reliability/security of distribution system network under supporting by distributed generation. Insight-Energy Sci 2(1):1–14

25. Eltamaly AM, Alolah AI, Abdel-Rahman MH (2011) Improved simulation strategy for DFIG in wind energy applications. Int Rev Model Simul 4(2)

26. Eltamaly AM, Khan AA (2011) Investigation of DC link capacitor failures in DFIG based wind energy conversion system. Trends Electr Eng 1, 12–21

27. Eltamaly AM (2012) A novel harmonic reduction technique for controlled converter by third harmonic current injection. Electr Power Syst Res 91, 104–112

28. Eltamaly AM, Alolah AI, Farh HM, Arman H (2013) Maximum power extraction from utility-interfaced wind turbines. New Dev Renew Energy 159–192

29. Eltamaly AM, Mohamed MA (2014) A novel design and optimization software for autonomous PV/wind/battery hybrid power systems. Math Probl Eng 2014

30. Eltamaly AM, Al-Shamma'a AA (2016) Optimal configuration for isolated hybrid renewable energy systems. J Renew Sustain Energy 8(4):045502

31. Narasimha Raju VSN et al (2019) A novel approach for reactive power compensation in hybrid wind-battery system using distribution static compensator. Int J Hydrog Energy 44(51):27907–27920, 22 Oct 2019. https://doi.org/10.1016/j.ijhydene.2019.08.261

32. Eltamaly AM, Farh HM (2015) Smart maximum power extraction for wind energy systems. In: 2015 IEEE international conference on smart energy grid engineering (SEGE), pp 1–6. IEEE

33. Shouxiang W et al (2006) Power flow analysis of distribution network containing wind power generators. Power Syst Technol 30:42–45

34. Eltamaly AM, Elghaffar ANA (2017) Load flow analysis by gauss-seidel method; a survey. Int J Mechatron, Electr Comput Technol (IJMEC), PISSN (2017), 2411–6173

35. Khaled U, Eltamaly AM, Beroual A (2017) Optimal power flow using particle swarm optimization of renewable hybrid distributed generation. Energies 10(7):1013

36. Eltamaly AM, Sayed Y, Elghaffar ANA (2017) Power flow control for distribution generator in egypt using facts devices. Acta Tech Corviniensis-Bull Eng 10(2)

37. Eltamaly AM, Al-Saud MS (2018) Nested multi-objective PSO for optimal allocation and sizing of renewable energy distributed generation. J Renew Sustain Energy 10(3):035302

38. Eltamaly AM, Elghaffar ANA (2017) Modeling of distance protection logic for out-of-step condition in power system. Electr Eng 11

39. Jamil E et al (2019) Power quality improvement of distribution system with photovoltaic and permanent magnet synchronous generator based renewable energy farm using static synchronous compensator. Sustain Energy Technol Assess 35:98–116. https://doi.org/10.1016/j.seta.2019.06.006

40. Eltamaly AM (2007) Modeling of wind turbine driving permanent magnet generator with maximum power point tracking system. J King Saud Univ-Eng Sci 19(2):223–236

41. Eltamaly AM (2017) Harmonic injection scheme for harmonic reduction of three-phase controlled converters. IET Power Electron 11(1):110–119

42. Eltamaly AM, Alolah AI, Abdel-Rahman MH (2010) Modified DFIG control strategy for wind energy applications. In: SPEEDAM 2010, pp 653–658. IEEE

43. Li L et al (2015) Maximum power point tracking of wind turbine based on optimal power curve detection under variable wind speed. Int Conf RPG 2015. https://doi.org/10.1049/cp.2015.0492

44. Camm EH, Behnke MR et al (2009) Reactive power compensation for wind power plants. IEEE Power Energy Soc Gen Meet, Canda. https://doi.org/10.1109/PES.2009.5275328

45. Mihet-Popa L et al (2010) Dynamic modeling, simulation and control strategies for 2 Mw wind generating systems. Int Rev Model Simul

46. El-Tamaly AM, El-Tamaly HH, Cengelci E, Enjeti PN, Muljadi E (1999) Low cost PWM converter for utility interface of variable speed wind turbine generators. In: 1999 Conference proceedings (Cat. No. 99CH36285) APEC'99. Fourteenth annual applied power electronics conference and exposition, vol 2, pp 889–895. IEEE

47. Rana AJ et al (2016) Application of unit template algorithm for voltage sag mitigation in distribution line using D-STATCOM. In: 2016 International conference on energy efficient technologies for sustainability (ICEETS), India. https://doi.org/10.1109/ICEETS.2016.758 3849

48. Laka A, Barrena JA, Zabalza JC, Vidal R (2013) Parallelization of two three-phase converters by using coupled inductors built on a single magnetic. Prz Elektotechnicy 89:194–198

49. Laka A, Barrena JA, Zabalza JC, Vidal R, Izurza-Moreno P (2014) Isolated double-twin VSC topology using three-phase IPTs for high-power applications. IEEE Trans Power Electron 29(57):61–69

50. Mehouachi I et al (2019) Design of a high-power D-STATCOM based on the isolated dual-converter topology. Electr Power Energy Syst 106:401–410

51. Matsui K, Murai Y, Watanabe M, Kaneko M, Ueda F (1993) A pulse width-modulated inverter with parallel connected transistors using current-sharing reactors. IEEE Trans Power Electron 8:186–191

52. Zhang D, Wang F, Burgos R, Lai R, Thacker T, Boroyevich D (2008) Interleaving impact on harmonic current in dc and ac passive components of paralleled three-phase voltage-source converters. In: 23th annual IEEE applied power electronics conference and exposition

53. Farh HM, Eltamaly AM (2013) Fuzzy logic control of wind energy conversion system. J Renew Sustain Energy 5(2):023125

54. Singh B et al (2018) GA for enhancement of system performance by DG incorporated with D-STATCOM in distribution power networks. J Electr Syst Inf Technol 5:388–426. https://doi.org/10.1016/j.jesit.2018.02.005

55. Mahela OP et al (2016) Power quality improvement in distribution network using DSTATCOM with battery energy storage system. Electr Power Energy Syst 83, 229–240. https://doi.org/10.1016/j.ijepes.2016.04.011

Wind Power Plants Control Systems Based on SCADA System

Khairy Sayed, Ahmed G. Abo-Khalil, and Ali M. Eltamaly

Abstract The objective of this chapter is to introduce the state of the art technology in wind power plant control and automation. This chapter starts with a historical background about supervisory control and automation evolution in the last decades. Several remarks are made regarding the use of SCADA Systems in wind turbine power plants. The Supervisory Control and Data Acquisition (SCADA) systems are responsible for controlling and monitoring many of the processes that make life in the industrial world possible, such as power distribution, oil flow, communications, and many more. In this chapter, an overview of SCADA at the wind power plant is presented, and operational concerns are addressed and examined. Notes on future trends will be provided. Finally, recommendations are provided regarding SCADA systems and their application in the wind power plant environment. One of the most significant aspects of SCADA is its ability to evolve with the ever-changing face of Information Technology (IT) systems.

K. Sayed
Faculty of Engineering, Sohag University, Sohag 82524, Egypt

A. G. Abo-Khalil
Department of Electrical Engineering, College of Engineering, Majmaah University, Almajmaah 11952, Saudi Arabia

Department of Electrical Engineering, College of Engineering, Assuit University, Assuit 71515, Egypt

A. G. Abo-Khalil
e-mail: a.abokhalil@mu.edu.sa

A. M. Eltamaly (✉)
Sustainable Energy Technologies Center, King Saud University, Riyadh 11421, Saudi Arabia
e-mail: eltamaly@ksu.edu.sa

Electrical Engineering Department, Mansoura University, Mansoura 35516, Egypt

K.A. CARE Energy Research and Innovation Center, Riyadh 11451, Saudi Arabia

Abbreviations

ANNs	Artificial neural networks
ANFIS	Adaptive neuro-fuzzy inference systems
CCTV	The closed-circuit television
CM	Condition monitoring
CMS	Condition monitoring system
DCS	Distributed control system
DC	Direct current
EPON	Ethernet Passive Optical Network
FIS	Fuzzy inference system
HVDC	High-voltage direct–current
HMI	Human Machine Interface
HV-IGBT	High voltage isolated gate bipolar transistor
IEDs	Intelligent electronic devices
IT	Information technology
KPIs	Key Performance Indicators
LVRT	Low Voltage Ride-Through
MPH	Mile per hour
MV	Medium voltage
NNs	Neural networks
OLT	Optical Line Terminal
OPC	Open connectivity
ONU	Optical network unit
PCC	PC Controller
PLC	Programmable Logic Controller
POS	Passive optical splitter
RTU	Remote Terminal unit
RPM	Revolution per minute
SCADA	Supervisory Control and Data Acquisition
SOE	Sequence of events
SVM	Support Vector Machine
TMA	VAWT vertical axes wind turbine
WTG	Wind turbine generator
WTC	Wind turbine controller
WPP	Wind power farms
WPP	Wind power plant
WTGs	Wind turbine generation system

1 Introduction

SCADA is an abbreviation that refers to "Supervisory Control and Data Acquisition." It is an essential tool to control and monitor various measurements of the wind turbine generation system (WTGs), and it's usual to include it together with the wind turbines. SCADA serves as the primary interface between the wind power plant operator and the wind farm equipment [1–4]. It allows integrating all the info about WTGs, meteorological mast, and substation in a single point of control, recapturing, and storing operation data from the WTGs and various alarm signals. Moreover, SCADA enables sending control signals from the wind power plant operator to the wind turbine controller. At present, with the trend toward electricity generation from renewable energy resources are widespread in many projects. Wind farms (WPPs) will be built in near future. This expected growing in wind power plants (WPP) will significantly affect control, operation, and control electricity network today. Many research work and investigations are conducted to study wind farm (wind) subsystems (Turbine, collector system, substations, etc.) that were discussed and addressed in many publications [5–10]. However, the infrastructure of SCADA systems and the related communication networks in wind power plants are relatively less processed and rarely discussed [10–12]. Typical wind power plant consists of wind turbines, meteorological system, and local wind turbine network, collecting point, and transformers substation. Power cables are used with various cross section areas to transfer power from wind turbines that are connected to the facility system through transformers and distribution lines [13].

The essential characteristics of SCADA required in a wind farm system can be summarized as follows:

- To integrate WTGs, substation, and meteorological tower data in a single system.
- To enable a frequent access to the wind power data from the local PC placed in the substation building and remotely from control centers.
- To enable modifying different control parameters of wind turbines.
- Communication protocols utilized by the system must be compatible with the others.
- Usually, to guarantee the safety of technicians working inside the wind power plant, a clear hierarchy must be predefined for all the users.
- Several parameters for each wind plant component must be displayed. Generally, the following conditions are required:

  1. WTGs data: digital data such as active alarms and status (ready, working, stopped, paused). Analog data such as power (kW), power factor, speed (wind turbine, generator, rotor), temperatures, currents, and voltages for the three-phases.
  2. Measurement tower: or Met (meteorological) Mast that measures wind speed and direction, pressure, temperature, battery status. A measurement tower or met mast can be a free-standing tower or a movable mast, which holds measuring instruments or meteorological instruments. Such instruments are

used to measure wind speed, wind direction, and thermometers. Met masts are important in the development of wind farms, as exact knowledge of the wind speed is basic to know how much energy will be generated, and whether the turbines will endure on the site.

3. Substation: Line voltages and currents, delivered active and reactive power, circuit breaker status, substation alarms, and protection system events. The user must be capable of changing various system parameters at any time. This includes also the opening and closing orders of the main switch.

4. WTGs: The start and stop commands of the WTG, use of the orientation system, transferring power generation data.

However, different reports can be produced with the data provided by the SCADA system, such as generated power, determination of the power curve, availability of the turbine, failures statistics, wind data (speed and turbulence), active, and reactive power and power factor (cos φ) at the substation. SCADA systems store, retrieve, and exports massive amount of data to a variety of stakeholders, everyone with diverse needs:

Remote control center: The control center should be able to manipulate the alarms and system conditions in a quick and effective manner, discriminating the root cause of faults without being concealed by cascading alarms. The SCADA system architecture for wind farms is shown in Fig. 1.

Remote monitoring and diagnosis: the remote operating center must be able to manipulate and interpret data rapidly to solve the operating and system problems.

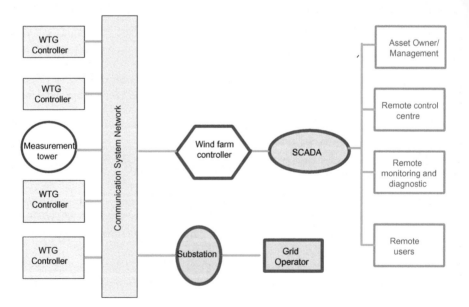

Fig. 1 SCADA system architecture in wind farm

Moreover, historical SCADA data can be used to authenticate computational models or improve new models.

Asset owners: This package uses SCADA output to calculate power revenue and calculation of energy losses, etc.

2 Wind Farm SCADA System

The wind turbine generator consists of the same wind turbines, circuit breakers, and step-up transformer. The generation voltage is stepped-up for each wind turbine using a voltage converter. Wind turbines are divided into groups, and each group is connected to a collector bus through the circuit breaker. Multiple grouped feeders are connected to a high-voltage (HV) transformer that raises the voltage to the transmission level [14–16]. Figure 2 shows a typical single line diagram of WPP. Wind power plants are divided into different regions: wind turbine area, collector feeding area, collector bus area, high-voltage transformer area, and transmission line area.

The main components of the wind farm are wind turbines, meteorological system, and electrical system [15]. However, SCADA systems are helpful in remote monitoring, data acquisition, data logging, and real-time control [16]. Remotely collect operation information from wind farm components and based on the information collected, the control center performs the appropriate procedures. Each WPP has a dedicated connection to the local control center for real-time monitoring and control. However, one control center can manage and control one or more wind power plants

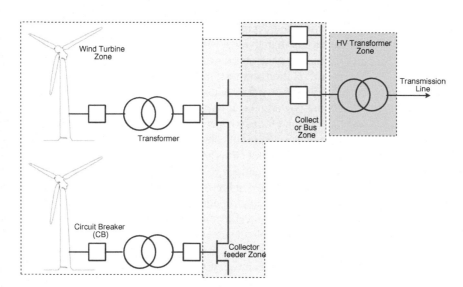

Fig. 2 Configuration of a typical wind power plant

remotely. There are many applications covered by SCADA systems in WPP. The three major applications are SCADA turbine system, SCADA wind power plant system, and SCADA security system [6–8].

Software system includes two types of software which are SCADA and applications software. The operator workstations are tied to networking architecture to enable monitoring operations in real-time. SCADA solutions include:

- SCADA solution to supervise, monitor, control, and report wind farm operations
- Sequence of events (SOE) recording for accurate alarm and event sequences that decrease troubleshooting time
- Network solutions include services that help with assessment, design, implementation, management, and audits of infrastructure
- Wind power application software
- Design and Operations Software that enables system expansion and upgrading
- Firewall and security software for Industrial Networks
- Wind Farm Management and Control.

A variety of solutions are required to effectively manage a wind farm. A high-performance wind turbine control system comprises SCADA software for monitoring, data acquisition, controlling, and reporting for wind turbine generators. Reliable automation systems and network technology support wind farms to fulfill with growing grid code regulations. SCADA systems and reliable wind turbine control systems enhance operation in wind turbine generator or totally in the entire wind farm. Generally, Wind turbine control systems are designed to maximize availability. Wind farms must respond quickly to the volatility of demand. A reliable control solution is required to optimize operations on individual wind farms and manage an entire fleet to increase efficiency, save costs, and improve overall asset management. The control system, together with the integrated wind turbine control unit and SCADA technology, can help manage both individual wind turbines and the wider wind farm resources to help reduce turbine generator downtime and increase availability.

The wind turbine control solutions embrace automation systems for wind turbines and wind farms. A broad range of wind turbine control systems can be used for offshore and/or on-shore wind power generation and wind farm management. These solutions assist wind turbines and farms to operate smoothly and cost-effectively.

Energy forecasts from wind farms are collected by system operators to improve the forward transmission schedule for wind farms and traditional generators to maintain system security with wind power fluctuations [14, 15]. The overall control framework of the wind power system is exemplified in Fig. 3. The wind farm control center takes power dispatch commands from the system operator. Consequently, distributes power reference levels to individual wind generator controllers, which in turn facilitates the wind farm to keep output power within the dispatch order from the system operator [16–19]. Furthermore, wind farms with power control competency are capable to contribute in the initial stage of wind power system restoration.

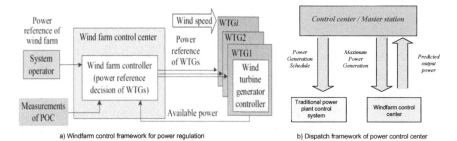

a) Windfarm control framework for power regulation

b) Dispatch framework of power control center

Fig. 3 Overall control framework for power system and wind farm

2.1 Wind Farm SCADA System Characteristics

Wind is an infinitely renewable energy source that can be harnessed as a premium energy source, given the right location and the latest turbine generation technology. On the farm, individual wind turbine generators are connected to the medium voltage (MV) gathering system and the communication network [20]. This medium voltage electricity then stepped up using a transformer to a high-voltage (HV) transmission system and electrical network. As the number of WTG generators increases within a wind farm, the need to manage these assets gradually becomes more significant.

The SCADA system offers real-time access to wind turbine generator diagnostics and generators and allows easy wind energy data management and continuous communication with remote wind power generation sites. Therefore, these systems should support multiple communication networks (microwave, cellular, fiber-optics network, radio, and more) and includes redundancy and failover schemes. Managing wind farm generating resources helps the reduction of turbine generation downtime and increases availability by making the wind SCADA system as a part of wind automation strategy.

2.2 Control of Wind Generation System Using SCADA

SCADA is used for supervision, monitoring, and control of wind turbines and wind parks remotely. The SCADA provides a full remote control and supervision of the entire wind park and individual wind turbines [21–23]. The SCADA system can run on the operator workstation in the control room of the wind power plant or it can be displayed on any internet-connected computer accessing the wind farm using TCP/IP communication protocol [24].

The overall control system of wind power plant is shown in Fig. 4. The main functions of the SCADA system can be summarized as follows:

- Wind park overview
- Wind park control

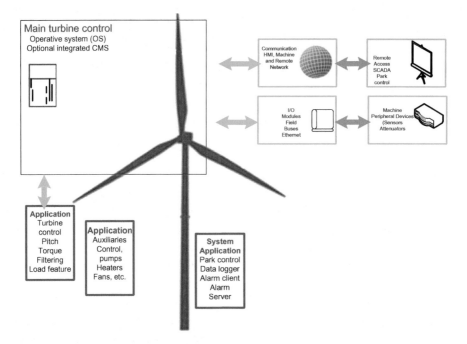

Fig. 4 Overall wind turbine control system

- Turbine overview
- Turbine control
- Log viewer
- Report Generation.

2.2.1 SCADA Systems in Wind Park

The SCADA system provides an overview of the wind park with a graphical user interface of the wind park representing the status of each individual turbine. Furthermore, currents, voltages, and wind power production data are shown.

2.2.2 Wind Park Control

The SCADA system in the wind park overview enables start/stop actions in the entire wind farm, groups of turbines, or separate wind turbines. Moreover, the park control is used for setting energy production limits for the entire wind farm. The aim of wind park control is to maximize energy production for the wind farm while reducing infrastructure and operating costs. For most projects, the economy is more sensitive to energy production change than infrastructure capital costs. Therefore, it is appropriate to use energy production as a dominant planning design parameter.

However, the complete design of the wind farm is assisted using wind farm design tools.

2.2.3 Wind Turbine Overview

The SCADA system has functions for turbine overview that gives a full overview of all important parameters of the wind turbine, for example, electrical parameters, rotation speed, pitch angle, temperature, and yaw system, etc. SCADA system is utilized to monitor the turbine parameter from the remote terminal area. Wind farm is controlled via a central interface that gathers the data from individual outstation components.

2.2.4 Log Viewer

SCADA event logger provides a complete high-level view on the power generation process that is continuous over time and captures information about user activities, system changes in the wind parks as well as system status updates. The SCADA system browses flexible structures of the log data of the wind turbine. All relevant log data are time-stamped, accessible, and can be arranged by various parameters.

2.2.5 Report Generation

SCADA system creates a report based on an alarm, status triggers create a report when something is on or off. Most owners require information that is regularly taken out from SCADA for operation analysis goals (loading studies, performance analysis, energy saving initiatives, etc.) [25, 26]. Reports can be generated automatically or manually. The reports can be generated by extracting information, and reformat usually a one-time time exceeds the cost to allocate and automate SCADA report. The report generator of the SCADA system makes it possible to make all relevant reports based on the data logging. The generated reports can be graphically represented to enable the best possible data review.

2.3 Main Tasks of Wind Turbine Control System

Figure 5 shows the wind generation control system. However, the main tasks of the wind turbine control system can be summarized as follows:

- Operational management and monitoring
- Wind park diagnostics, safety
- Communication, reporting, and data logging.

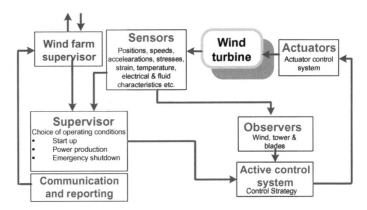

Fig. 5 Wind turbine control system architecture

Operational states of wind turbine comprise the following states: Idling, Start Up, Normal power production, Normal shutdown, and Emergency shutdown [15, 27–35].

The main input data to SCADA system include Wind speed, Rotor speed, Blade pitch, Electrical power, and Temperatures in critical area. However, some systems include strains, stresses (tower, blades), speed, position (yaw, blades, actuators, rotor tilt, teetering angle, fluid properties and levels, electrical system (currents, voltages, utility grid characteristics), icing conditions, lighting, humidity).

2.4 Wind Farm SCADA System Functionality

The SCADA system displays the WTG components in a graphical user interface and records the wind turbine conditions automatically. The system records the turbine availability, events, power generation, and fault data in real-time. One can use the pan, tilt, and zoom functionalities in the display panel. Therefore, wind farm operators can click on the figure representing wind turbine and monitor turbine performance by accessing a set of real-time data. Generally, the SCADA system in wind farm compromises seamless and open connectivity (OPC) and the operator can access dashboards anywhere, anytime, and on any connected device.

Wind farm SCADA system enables users to monitor and control remote operations and turbines in real-time. Operators can access precise, real-time information including updated weather and meteorological informs as well as fully configurable Key Performance Indicators (KPIs). Comparing energy production reports against historical information helps toward achieving accurate forecasts. However the master-station network is shown in Fig. 6.

Wind farm SCADA systems provide users with rich SCADA visualization and reports that are integrated real-time. Consequently, historical geographical terrain maps, and a quick overview of multiple operations and plants located anywhere in

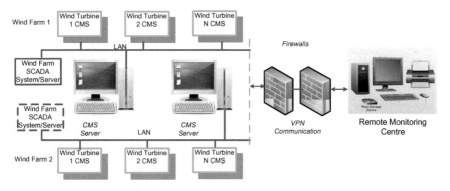

Fig. 6 Master-station network

the related electrical network. The SCADA wind farm system software is deployed over the web, is published, customized across the web, and delivered to any platform via different software technology [36, 37].

Based on open connectivity via OPC, Modbus, BACnet and IEC61400-25, the wind SCADA system software permits the integration of equipment, process and professional data into a single integrated plant operations' view, and provides control actions of the wind farm operation. Users are provided with a reliable, safe, and immediate response to energy, environment, and work requirements. SCADA in wind farm combines scalable mapping technology with classic supervisory control and data acquisition. For geographically dispersed assets, SCADA provides safe, scalable, and safe visualization and tracking of assets and wind turbines via GPS coordinates. SCADA technology enables easy navigation to quickly display alarm conditions and the condition of any site worldwide. Within seconds, the performance of assets, turbines, problems, or alarm conditions can be located and determined by integration with application software.

3   Structure of SCADA for Wind Power Plant

A SCADA system is responsible for gathering and managing all data collected from the remote terminal unit (RTU) at the outstation wind turbine and the SCADA server at the control center master-station for the sake of monitoring and supervisory control. The basic communication in the SCADA system utilized with WPPs is illustrated in Fig. 7. However, all the data from the wind farm are collected and sent over the communication link such as optical-fiber cables to the master-station (control center). The SCADA server in the WPP is an industrial computer which is considered as masters in the SCADA system while all RTUs act as slaves [1]. Sometimes, one RTU acts as a master to collect information from slave RTUs.

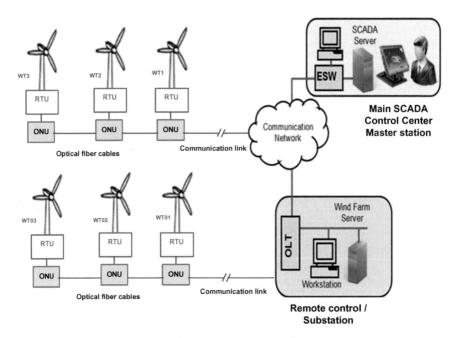

Fig. 7 Communication system architecture in wind farm SCADA

Due to increasing numbers of wind farms and its related components, it should be integrated into smart microgrid. These components are secondary power substations, distributed energy resources, public charging stations, virtual power stations, microgrids. However, new energy automation applications, such as automatic meter reading, meter data management or demand response, lead to extra communication requirements.

The communication networks for wind farms systems should be designed carefully. In this respect, the development of cost-effective and reliable communication architecture is vital. In this regard, Ethernet Passive Optical Network (EPON) is considered as a promising and attractive option. Fiber-optic based communication networks are utilized for advancing automation solutions. The network consists of an optical network unit (ONU) that is deployed on the wind turbine site. All ONUs servers wind turbines are linked to a main Optical Line Terminal (OLT). The EPON architecture of the OLT unit and the ONU unit placed at wind turbine outstation are comparable to the master/slave operation between the SCADA server and the RTUs in the WPP. The SCADA server polls data from RTUs at well-defined time-periods, and it can conduct control actions as needed. Therefore, in EPON system the OLT polls data from ONUs every few milliseconds. To explain this theory, a simple data-polling scheme is illustrated in Fig. 7. For instance, consider a polling cycle time to be 2 ms, this means that the wind turbine will be scanned for data 500 times during 1 s. In this arrangement, the OLT control center sends a GATE message to WT_i, and then waits for data until the control center sends it to WT_{i+1}. In this arrangement, an effective improvement in the performance of the OLT network could be achieved in terms of communication channel utilization and average delay of packet.

4 Data Network Configuration for SCADA System

The SCADA system is a vital component of this process. All wind turbines have a control box on top that contains PLC or RTU, power adapter, control panels, and I/O. Data of wind speed, wind direction, shaft rotation sensors are collected and transferred to the PLC. After sensing the wind direction, the SCADA control system can utilize the yaw gear motor to convert the turbine completely in the right direction to track the maximum generated power. All RTUs (or PLC) related to wind turbines are connected to a local area network (LAN), where the control box in each antenna uses an Ethernet network to be connected to redundant fibrous LAN link that is fixed at the bottom of the tower. The local network LAN is connected to a remote control master-station that operates the control system that gathers and manages data, regulates turbine settings, troubleshooting, offers intelligent alarms and reporting functionalities in the master-station.

These individual turbines, substations, meteorological stations, and other wildlife monitoring systems are connected to the central control room in Wind Control Center. It provides visibility to the operator to oversee the behavior of all wind turbines on all wind farms. By maintaining a log of activity on an interval basis, SCADA enables the operator to define corrective and corrections actions, if any, to be engaged. The system records the output power, availability, events, and alarm signals. It provides the ability to implement various control requirements in the voltage drop, power factor, and interactive energy generation. Therefore, the wind power plant contributions to both the voltage and frequency of network are facilitated. Operator workstation enables operators to manage the output power according to network requirements in real-time. SCADA communicates with the turbines over a communication link that uses optical fibers for almost all of its bonds. Wind turbines of various types can be controlled by one SCADA system. Some turbine suppliers provide their control/HMI display system.

The main advantages of SCADA system are that it can be used for different types of wind turbine. The PLC can provide data reports and analysis formats regardless of the type of turbine. It is important for wind farm operators to use many types of turbines and countless PLC types. SCADA should be easy to use and easy to configure. The ability of demonstrating animated mimics, using of pop-ups, and reduced risk of overlapping important information demonstrate simplify of the SCADA software. Also, creating content and behavior templates ensure the consistency of all the animations in the display boards. SCADA uses the rights to access and multi-level menus that are associated with each operator or user. Therefore, navigation within the application is tailored to the permissions and needs of each individual user. This ensures a layer of security, tracking, and control of user actions.

Wind turbines are installed to operate in coordination with other sources such as nuclear, solar, and hydro power in a network arrangement to improve performance. At present, congestion has become a major problem as wind energy suppliers balance energy production with the inputs available for transmission. A more scalable/modular system is required to accommodate the recent trends of renewable energy market.

To manage their growing industry, fiber-optic networks were installed on wind farms. Generally, there is a central facility whereby the SCADA system is capable to access wind farms throughout the country remotely and to access station alarms, events, and conditions. Centralized SCADA configuration offers the capabilities for traceability and management of the various wind application. It also updates the stations events automatically. Generally, each wind turbine gives about 300–350 data points.

SCADA alarms are highly configurable to address the diverse requirements of maintaining wind farm application. Alarm messages can be displayed, printed, and organized in alarm lists and archived.

Moreover, WPP operators can configure alarms using groups, sorting, acknowledgment, filters, and hiding. It also creates alarm meters and links specific actions to any alarm. Operators can recognize alarms directly from the mimics and these actions can be automatically broadcasted to all workstation on the Master-station computer network.

SCADA in wind farms uses OPC as the communications protocol, besides other protocols to collect data from different types of PLCs/RTUs. Wind power plant application often uses OPC and a driver to communicate smoothly with miscellaneous systems. OPC data access client exchanges data with communication servers in realtime. Moreover, the OPC data acquisition server assists data interchange with third party applications. The collected data are routed back to the master-station of control center. SCADA software in main control center should be proven to be high functional operation and user-friendly. Moreover, it should prove scalable, reliable, and easy to configure.

Wind SCADA provides a single user view that provides easy visual viewing and comprehensive management of turbine-equipped control systems. A simple and easy system is required to read graphical user interface so that can interact automatically to monitor the weather, managing, and controlling the turbines. The SCADA software communicates with the wind turbines through the graphical user interface (GUI) that acts as a light client for the application program and managing data elements. This configuration provides the user/operator with all the essential information about the turbine signals.

SCADA uses the distributed client–server architecture with the iteration mechanism to ensure that the design bears errors. By using the included redundancy features, SCADA is able to ensure continuous data collection in the event of a system component failure. Moreover, it also supports dual networks redundancy for communication with field equipment and between wind turbine stations, see Fig. 8. Every component in any substation in the configuration has a validity status to allow operators to display the system status in real-time. These OPC client stations communicate with

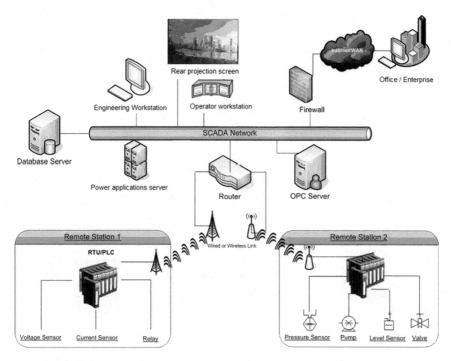

Fig. 8 SCADA system network configuration

increased front ends connected to a 1000 Mbps TCP/IP Ethernet network. Each front end is capable of receiving up to 60,000 input/output. The SCADA server communicates with the turbines via a communications network. The communication method depends on the distance from the master-station. Usually, WPP SCADA uses optical fibers to communicate with the nearby wind turbines.

Using the SCADA architecture, WPP operators can see in-depth details of data gathered from remote-site wind turbines in a real-time. WPP supervision is organized on two levels to manage the large amount of data and to enable the operation and troubleshooting of the facilities.

The first supervisory level offers an overview of the most important warnings, values, and counters, which are sufficient to oversee normal turbines and detect faults that need to be corrected. A more detailed supervision level is turned on demand to display the specific data from the turbine so that operators can instantly diagnose any malfunctions that occurred and accurately determine the treatment operations. Received data can be processed as defined points, historical reports, alarm management, and data trends, etc.

The control system collects all major operational information from outstations, generators, and associated substations. The control system is connected to the master-station control center through a remote communication channel, which facilitates maintenance. The center receives and processes this information in a streamlined and simplified structure that allows for easy identification and diagnosis of failure. This triggers the appropriate procedures to resolve it: remotely reset or activate local maintenance teams. As a result, the average downtime reduces and availability of the system increases.

5   SCADA System Instruments

The SCADA system in WPP connects the individual turbines, the wind power substation, and meteorological stations to a central master-station. The associated communications system allows the WPP operator to monitors the performance of all the wind turbines and also supervises the wind farm as a whole. The system keeps a record of all events and activities that allow the operator to determine necessary corrective actions. SCADA instruments record the produced energy, output availability, and error signals. The SCADA system deals with instruments to measure reactive power, voltage, and frequency. WPP SCADA dispatches generated power according to instructions from the network operator (regional control center).

In addition to the basic equipment needed for a working wind farm, it is also desirable, if the size of the project can guarantee investment, to erect some permanent meteorological devices on opposite masts. This equipment allows carefully monitoring and understanding the performance of the wind farm. In the absence of good wind data on the site, this decision will not be possible. Usually, large wind power plants contain one or more permanent meteorological masts, which are installed during the installation of wind power plant.

6   Wind Energy Power Plant Management System

At present, there is a strong focus on designing planning and operating tools for operating the electrical system under random demand and production conditions (which are still well anticipated). The system has complete control only over wind park management, and the overall power system may not operate in an improved manner. Therefore, systems that combine energy management systems at the transport and distribution level should be considered.

Smart grid technologies, along with forecasting tools, warehousing facilities, and demand-side management, may lead to new opportunities for increased integration capabilities. In this context, wind farm output forecasting tools within a maximum of 48 h should be more accurate than the current ones. In addition, both large-scale and regional projections of advanced network and energy management systems will be required. Besides the need to improve the physical integration of wind energy, management tools, and systems will be needed to integrate random wind energy into existing energy markets. Once again, advanced forecasting systems combined with improved storage capabilities and management measures are key to achieving this goal. Then the technical and economic aspects of the systems will be considered, in order to enable the incorporation of wind energy production on the grid and into the electricity market. The power management system is designed for distributed wind power system; the power management system switches the power supply mode and controls the system according to the wind power condition and load requirements.

A distribution generating system, along with a battery bank, can provide the user with reliable electrical power as shown in Fig. 9. The power management system is implemented from the microprocessor and data acquisition system. This power management system is applied in experimental equipment. A hybrid generating system controlled by the power management system, when random wind speed and solar radiation difference appears, provides constant electric power. A wind turbine management system that regulates the output power of the turbine, where the wind turbine includes a rotor with at least one rotating blade set at an adjustable angle to the rotor and the rotor management system regulates the speed within the predetermined wind speed range by changing the angle of the rotor blade in order to adjust Nominal output and reduces, by increasing a specific threshold that depends on the wind speed, the output power of the operating turbine to a smaller amount of nominal output but greater than the output where the wind turbine is turned off, and where the threshold value dependent on the specific wind speed is the angle of determining the specific rotor blade.

The management system maintains the angle of the rotor blade at a constant value until the nominal output is reached and the management system adjusts the angle of

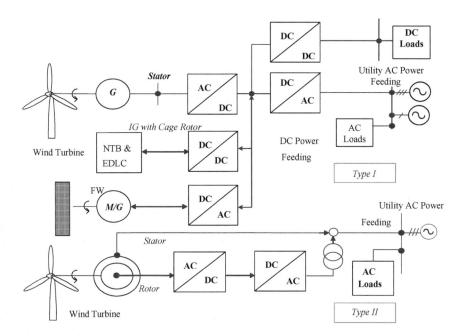

Fig. 9 Distributed wind power system

the rotor blade with respect to wind speed in order to maintain the nominal output at a fixed value.

A method for regulating the output power of a wind turbine with at least one rotating blade set at an adjustable rotor angle that includes steps: regulating the rotor speed within the predetermined wind speed range by changing the angle of the rotor blade in order to adjust the nominal output; The operational turbine output, which exceeds a specific threshold value dependent on wind speed, is reduced to a smaller amount of nominal output but greater than the output where the wind turbine is turned off, where the specified wind—the velocity-dependent threshold value is a specific angle for the rotor blade.

The offshore installation is one of the leading trends in wind turbine technology. There are significant wind resources at sea to install wind turbines in several areas where the sea is relatively shallow. Marine wind turbines may have a slightly better energy balance than land turbines, depending on wind speed conditions. In places where wild wind turbines are usually placed on flat terrain, offshore wind turbines can commonly produce about 50% more energy than the turbines placed on a near land site. This is because there are no obstacles and less friction on the sea. In addition, building and laying the foundation needs 50% more energy than land turbines. However, it must be remembered that marine wind turbines have a longer average life than land turbines. The reason is that the low turbulence at sea gives less stress to wind turbines.

Therefore, offshore wind turbines require increased corrosion protection, while reducing maintenance and service requirements and improving the supervision and control system. Inner corrosion protection comes from improved coating systems and a dry environment inside the machine. Prerequisite for a dry indoor environment is a sealed machine. The gear and generator are cooled by heat exchangers that recycle the air used in the air cooling system, instead of the traditional air-cooled components in previous turbines. To maintain low indoor air humidity, dehumidifiers are placed in the tower and nacelle chamber. The dehumidification system maintains the internal relative humidity lower than any steel corrosion risk limit (60%). For additional protection, the main electrical components (generators, control systems, etc.) have backup heating systems, which prevent condensation, even during sudden temperature differences.

7   Standards of Grid-Connected Wind Farms

7.1   Voltage Fault Ride-Through Capabilities of Wind Turbines

To enable widespread application of wind energy without compromising the stability of the power system, the turbines must remain connected and contribute to the electrical network in the event of a disturbance such as a voltage drop. Wind farms must behave as conventional power plants, to supply the active, and reactive powers to restore frequency and voltage, immediately after the error occurs.

Wind turbines can now remain online, for the first time, supplying the reactive power to the electrical network through major system disturbances. The Low Voltage Ride Through (LVRT) feature enables wind turbines to meet transmission reliability standards similar to those required from thermal generators. LVRT adds significant new flexibility to wind farm operations while more facilities require it.

7.2   Power-Quality Issues in Grid-Connected WPPs

The reason for this interest is that wind turbines are potential sources of poor energy quality. Measurements show that the power-quality of wind turbines has improved in recent years. Especially, variable speed wind turbines have some flicker-related advantages. However, a new problem arose with the variable-speed wind turbines. Modern inverters with forced commutated function used in variable-speed wind turbines not only produce harmonics, but also harmonics.

The power transmission system with optional reactive power (VAR) control provides support and control for the local network voltage, which improves transmission efficiency, and provides an interactive utility power grid (VARs), which increases network stability. VAR function maintains specified network voltage levels and power quality in fractions of a second. This feature is especially useful with weak grids or larger turbine installations.

7.3   Variable Speed Control

Variable speed control is utilized to maximize wind power capture and reduce turbine-load group loads. The wind turbine control system, through its advanced electronic devices, constantly adjusts the angle of inclination of the wind turbine blade to enable it to achieve optimum rotational speed and maximum towing lift at every wind speed. The "variable speed" operation increases the capacity of the turbine to remain at the highest level of efficiency. In contrast, constant speed wind turbines achieve the highest efficiency at a single wind speed. The result: increased annual energy production compared to plants running at a constant speed.

In addition, while the fixed speed rotors must be designed to direct strong wind loads, variable speed operation enables the loads from the storm to be absorbed and converted into electrical energy. The torque of the generator is controlled by the frequency converter. This control strategy allows the turbine rotation to override in strong winds and gales, thereby reducing torque loads in the powertrain. Variable-speed wind turbines convert the extra energy in the wind gusts into electrical energy. The turbine speed range is noticeably broader than the "slip" range used by other technologies, which produces heat rather than electrical energy when regulating energy in strong winds.

7.4 Active Damping

The variable speed operation provides active damping of the entire wind farm system, which results in significantly lower oscillation of the towers compared to the fixed speed wind turbines. The active damping of the machine also reduces the maximum torque, which provides greater reliability for the powertrain, reduces maintenance cost, and increases the life of the turbines.

Active-damping of tower oscillations is accomplished using the pitch angle as the control input. The flexible multi-object system is used to derive a directed model for controlling the first bending mode of the tower, which works to design a stable control law. The oscillation damping is integrated with a scheduled feedback control for multivariate gain that allows tracing the paths required for angular velocities of both turbine and generator.

8 WPP Control System

This section explores variable speed operating schemes for wind turbine generation applications. The main goals are to maximize energy production, provide tight startup, and reduce torque load on system components. This is done while maintaining the control strategy that operates in the variable power generation mode between the cut-in speed up to the speed of the main generator 900 rpm and working in the continuous power operation mode from the base speed to the maximum speed of the 1350 RPM generator.

The main functions of the proposed control and management system are:

(1) Supervising and controlling the interconnection of the wind turbine power station with the utility network.
(2) Control the frequency converter output performance,
(3) Improve the energy conversion efficiency of wind turbines,
(4) Provide system performance measurements to evaluate the operator,
(5) Achieve a safe closure under normal and emergency circumstances.

The control strategy shown here uses the rotor speed, torque, and generator strength as the feedback signals. In the normal operating area, the rotor speed is used to calculate a target energy that corresponds to the optimal operation. With power as a control target, the power transformer and generator are controlled to track the target power at any rotational speed.

8.1 Wind Turbine Control

One of the main aims in the development of offshore wind energy is cost reduction and increasing the availability. Due to the larger turbine sizes load control is becoming more and more important. Wind turbine control is essential to guarantee low maintenance costs and efficient performance. The control system also ensures safe operation, optimizes output power, and ensures long structural life. The schematic diagram of control unit of wind turbine is shown in Fig. 10. Maintaining the system stability is becoming more difficult as the frequencies of the turbine are getting closer to each other as well as closer to the frequencies of wind and waves. The turbine control system therefore is of crucial importance. In research domain, models for the design of advanced wind turbine control algorithms were developed. The control system of wind turbine is illustrated in Fig. 11. Those models and tools are including aerodynamic and structural dynamic modules. With the control tools, multi-parameter control algorithms can be developed, taking into account the complex and strong dynamic influences to which the turbines are exposed. This approach offers solutions for the following specific operational problems:

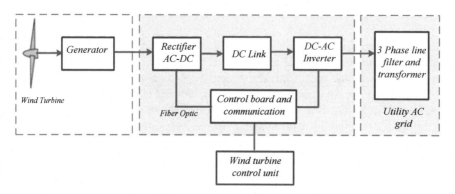

Fig. 10 A simple wind turbine control unit

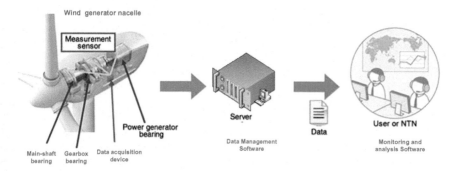

Fig. 11 Wind turbine Management system

- unnecessary stand-still due to an isolated approach of control and safety systems;
- high costs and limited possibilities for up-scaling due to high turbine loads and stability problems;
- uncertainty of energy output and high loads at extreme weather conditions;
- Accumulation of damage in case of a turbine shutdown caused by a severe failure.

8.2 Sustainable Control

The sustainable control is a developing and integrated design approach for the control system of offshore wind turbines. In this approach four parts can be distinguished:

- Optimized Feedback Control
- Fault-Tolerant Control
- Extreme Event Control
- Optimum Shutdown Control.

8.3 Design Tool for Wind Turbine Control Algorithms

Due to the increased importance, the increasing complexity and the increasing inter-action with the structural and aerodynamic design, knowledge, and tools which enables wind turbine manufacturers to develop and validate their own advanced control algorithms. An open-source design environment is important for developing and Real-Time testing of control systems. Figure 12 shows wind farm energy systems that are based on reliability and maturity of technology. Control models and strategies for wind farms have been developed, with the aim of improving the operation of wind farms taking into account participation in controlling the energy system (frequency)

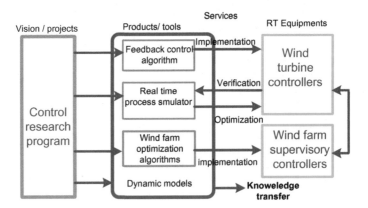

Fig. 12 An overview of the wind farm control

and reactive energy (voltage), maximizing energy production, improving the impact on quality energy and reducing mechanical loads and life-time consumption. Wind farm control will rely on signals from wind turbine controllers. These signals will be used to predict wind speed in individual wind turbines in the short term (i.e., from seconds to minutes) in order to enable individual wind turbine controllers to respond in an ideal manner to wind speed fluctuations.

The concepts for wind farms include electric components such as squirrel cage induction generator, doubly fed induction generator, and modern high-voltage direct current (HVDC) techniques. HVDC systems are based on converters with full control capabilities that are based on force commutated semiconductor devices. The included wind turbines will be pitch controlled as well as (active) stall controlled. Consequently, the characteristic curve of the torque speed is formed to output the generator to increase the energy conversion to the maximum power capture area. However, the target power is constantly updated at any rotation speed (rpm) in the operation range. During extreme operating condition, for example during startup, shutdown, generator overload, or overspeed, various strategies driven by other system considerations should be used as variables to control input as well. Wind energy is proportional to the cube of wind speed. Therefore, in order to regulate output power as the wind increases, there should be a mechanism to decrease the output power of the wind turbine while the capturing area remains constant. Wind turbines with constant speed automatically achieve this, because in high winds their blades stop. The resulting lower lift and increased drag significantly reduce the blade's ability to fine wind capacity. The vector control strategy can be used for frequency converters. By using the vector control, the specified torque and limitation can be effectively maintained. With the closed-loop control, maximum dynamic performance could be accomplished. This is because the current components of the torque and flow can be controlled independently of each other. However, closed-loop frequency control can be realized with a speed sensor or sensorless control. During normal operation of wind power stations, constant variations occur in both wind speed and direction. As the wind velocity increases, the turbine rotational speed will increase slowly. By adjusting the frequency of the frequency converter, the velocity of the stator field of rotation can be adjusted to yield the desired slip and thus the desired torque can be obtained. Therefore, the rotating speed of the wind turbine must adapt to the prevailing wind speed. If the wind speed of the wind turbine increases above the generator base speed, the frequency converter increases its frequency while decreasing the torque reference in order to achieve rated or constant output power up to the maximum rotational speed of the generator. If the wind speed drops below the starting speed or the traveled speed, the wind turbine will stop and the frequency converter will go into standby mode. Thus, by controlling the frequency, one can control the output power and torque of the wind generator. This results in optimum wind turbine performance over wind speeds and wind conditions. There are two different control strategies. The first is to control the generator speed in relation to wind speed and wind turbine rotational speed by providing a speed reference signal to the frequency converter. The second method is to control the generator torque with respect to wind speed

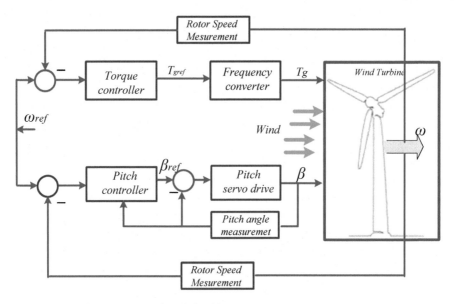

Fig. 13 Detailed control system for wind turbine

and wind turbine torque by providing a reference torque to the frequency converter. Figure 13 shows a simplified control strategy for wind turbine.

The favored control strategy is the generator that is controlled to provide the required torque and wind velocity relationship. The wind turbine can operate near the maximum power factor at low and moderate wind speeds, which represent furthermost of the energy capture at specific wind locations. In case of high wind speed, the torque is controlled to keep a steady energy production while increasing of WT generator speed. Moreover, the transition from moderate to high wind speeds should be occurring smoothly to avoid overloading the WTG components. The reason behind choosing the subsequent strategy as a preferred strategy is that there will be fewer fluctuations in energy on the grid in case of locations that have high wind speeds. As the wind turbine rotation decrease, the maximum force that can be generated reduces. At low and medium wind speeds, the generator torque is controlled to work near the optimum output curve producing the maximum energy. The rotor speed is expected to be regulated and limited within the higher limit of the generator RPM region although existing of higher wind speeds at the WPP site. The gearbox design and gear ratio selection are also critical in getting optimal operation.

9   Wind Power Plant Control and Management

Computer Console (PCC) is an automation engine that is designed to meet automation system and wind plant management requirements for substation applications. PCC is a platform for information control and automation applications including the human sub-machine interface (HMI), sub-station data, fault data extraction and indexing, sub-station level sequence collection (SOE), power quality analysis, predictive/preventive maintenance, Smart reports, and condition management. The PCC also delivers data management tools for treating historical and diagnostic data according to SOE.

The wind turbine controller consists of an external loop controller, which can be an RTU or PLC. The PLC is connected to the wind farm management system. The communication interface is implemented by a communication server or processor on the PCC end and connected to the PLC through a PROFIBUS-DP communication link. PROFIBUS DP is the distributed I/O protocol. It allows for ultra-high-speed periodic connectivity that includes small amounts of data with data transfer rates of up to 12 Mbps. PROFIBUS is the carrier system for communication in small cell networks and with the field device. PROFIBUS is European standard EN 50 170 and is applied worldwide in the environmental field. PROFIBUS provides the advantages of being a reliable and high-speed network that can easily interact with all smart subsystems used in the wind power plant. PROFIBUS can be either shielded by two wires of fiber or optical fibers. Optical fibers provide the advantage of use in the external environment while avoiding problems due to lightning. The plant control and management should ensure reliability and automated operation of WPPs. For achieving this, the related components and WT variables should be supervised and monitored. The supervision is done by continuously maintaining the value ranges and setting permissible values of system variables. The control and management decide predefined operating states and identify errors and emergency situations immediately. For this, the combined wind turbine frequency transformer, external loop control system (PLC), and factory management system (PCC) together should influence the wind power operating behavior based on pre-set control signals and required values, and interaction of changes in system variables or errors. In addition to reliable operation, another very important goal is to achieve the perfect match between the output power quality, the low mechanical pressure, and the electrical load of the plant on all its components. Power stations are equipped with frequency converters to convert wind turbine energy into electrical energy and provide consistent low-quality distortion energy to the utility grid. The malfunctions feedback will be determined based on operational conditions and control logic flow charts along with reaction agenda documents and attached with wind turbines.

10 Operation of the Outer-Loop Controller

The external-loop control system is implemented using the PLC controller. The structure of the program and all internal frequency converter functions and variables are monitored continuously along with many external device functions and variations. Depending on the results of the logical decisions, the method of operation and the points of determination of the specified value, the appropriate decision and reaction will be taken. Single wind turbines are normally powered in automatic mode. However, manual and semi-automatic operating methods with manual input of required values are necessary during commissioning, troubleshooting, and maintenance.

Temporary operating conditions may last only for a limited period. Thus its duration is monitored. After exceeding the preset maximum intervals, the error is turned off, as an error should be assumed. The duration of static operating cases (S) is not monitored by the control and management system. The manufacturer remains in these cases as long as all normal operating conditions are met. In all operating conditions, all normal operating conditions must be checked continuously. Only one condition is required to switch between "Stop operation," "Normal Shutdown," "Fault/Overspeed Shutdown," or "Emergency Stop" states. By contrast, to start the "start" or "Run-up" operating states, all conditions must be met.

10.1 Wind Farm Plant Testing

The monitoring components and the influencing and changing variables must be tested and recorded after successful commissioning of control and management systems. Outputs of all subsystems must be queried for shutdown values and all mechanical actuators for testing purposes. Sensors can check the correct reactions for configurations. If errors occur, they must be recorded. Errors cause the additional process to stop until the error is corrected and the factory release manually. All plant components and their marginal values should be checked in all operating conditions. This system works properly for all systems, operating temperatures, and ground error condition. After successful test, the factory goes to a later operating state; Otherwise, the factory operating state test is repeated until all release conditions are fulfilled, such as operator orders, unlock after emergency stop, network availability and appropriate function, component function, ground failure detection, temperatures and marginal values.

10.2 Stop Operation

Fixed rotor characterizes shutdown state. Moreover, in this operating condition, the rotor mechanical brakes are activated and can also be tested to obtain the appropriate functions. The generator is turned off and disconnected from the supply network. The conditions that prevent the manufacturer from moving to the initial operating state are tested. Then a system scan is performed. If all conditions are positive, start conditions are checked. If this is also achieved, the change to "start" will occur. As with factory testing, the appropriate operating condition number distinguishes "mechanical brake assembly" and "contact switch" for the generator open. The available wind speed is continuously monitored in each individual wind power generator independently by a wind meter, such as an anemometer. Turning on individual wind turbines occurs when minimum wind speeds are available.

10.3 Starting of Wind Generation System

The rotor is stable and still attached to mechanical brakes. In a repeated sequence, the conditions for turning off errors are tested and initiation of appropriate routines if necessary. The starting signal from the management system is also examined during this sequence. PLC gives a wired output signal to the digital control unit to start the frequency converter. The frequency converter then passes its power on the sequence routine. Bipolar Isolated Gate Transistors (IGBTs) are checked and feedback signals tested, cooling fans are turned on, the DC connection is pre-charged, then the frequency converter turns off the generator connector switch. Once the generator contactor switch closed status is achieved by the PLC, the wind turbine moves to the "Standby" operating condition, provided that the trigger order for the "Automatic" command is given by the WPP management system.

10.4 Standby State

In the event of readiness, all components of the wind turbine generator are constantly checked to determine whether they are indeed ready for operation (standby state). In a repeated sequence, the conditions for turning off errors are tested and initiation of appropriate routines if necessary. The average available wind speed is also constantly checked to determine the minimum available wind speed. The generator system is already connected to the network supply system during the previous operating state. Breakdown closure conditions, operator orders conditions, and operating settings are checked one by one. If the suitable conditions are met, the relevant operating cases are entered. If standby is kept for a long time, for example, five minutes or more, the management system is notified with a message that the wait time has been exceeded and the control goes to shutdown and then returns to the steady shutdown state.

10.5 Run-Up State

Speed can be increased if the combined average wind velocity is greater than 5 m/s and the instantaneous wind speed is not very high (Run-up is possible). The frequency converter is first checked to ensure that it is ready for power generation as well as the network connection system that is still connected. In repeated sequences, normal shutdown and shutdown conditions are tested and suitable procedures are initiated if necessary. The speed and torque references are checked for an initial value of zero and if this is correct, then the inverter gate will be enabled. Once the inverter gate is enabled, the frequency converter will begin capturing and synchronizing the rotary generator. The rotation direction and acceleration limit values are checked. The torque or velocity reference value is then released from the PLC and adjusted itself to match the feedback values of the charged generator rotor. If synchronization is achieved within the allowed time period, the rotor speed of the wind power plant can be operated to a value at which sufficient active energy can be generated for the grid. With little or no interactive force pulled from the network via the frequency converter, the rotor speed lies within a range determined by the outer ring control system, and is affected only by the available wind speed immediately. The wind turbine generator is now ready to go into a stable variable-power operating state.

10.6   WPP Variable Power Operation

When the desired target speed is achieved, the generator system and frequency converter are able to generate enough active energy for the supply network, then the electrical energy of the grid system is provided. In the variable energy process, the generator system provides variable electrical energy (kWh) in the supply network. The sliding frequency phase angle is adjusted or adjusted to the optimum value, so that maximum output power or minimum component loads are possible. The outer-loop controller provides a reference value for the output power related to wind speed and energy demand. In the variable power process, the speed or output power is regulated by the frequency transformer of the generator system. Torque or speed is maintained within a permissible range by adjusting torque or speed according to the previously required values. When the control reserve value is reached, the target torque or speed value is changed according to the characteristic line of optimum power speed. Throughout variable power operation, normal shutdown condition, malfunction shutdown conditions, and standby shutdown conditions are constantly tested. In the variable power process as well, all conditions of normal operation are checked, and if necessary, appropriate actions are initiated. It is important to check temperature limits, acceleration limits, vibration level, low-speed limits, and power stability limits periodically. Essential messages are listed in the Variable Power On mode. Due to the sufficiently high average available wind velocity, the wind turbine generator automatically switches to the "continuous power" steady state.

10.7   WPP Constant Power Operation

If the average wind velocity is available high enough, the wind turbines inside the power plant will move from "variable energy" to the "continuous energy" process. In this operating condition, the external loop control system determines the desired values of rated output power, nominal system output, and fluctuating range. Rotary values for speed/torque, slip frequency regulations and torque bounds are constantly supervised to adjust the constant output power. Thus the output fluctuations in the turbine lead to small changes in speed. Speed is maintained within the regulation range by lowering the torque and adjusting the slip frequency. A small set of overload is permitted in the event of instant wind gusts, so that the wind turbine speed does not need to be adjusted quickly or often. However, the overload range should be of limited duration, depending on the thermal behavior of the entire system. During continuous power operation, normal shutdown conditions, malfunctions/excessive shutdown conditions, and "standby shutdown" conditions are constantly tested. In continuous power operation as well, all normal operating conditions are checked, and if necessary, appropriate procedures are initiated. The vibration level, acceleration limits, temperature limits, over-speed limits, and power stability limits are constantly checked. Figure 14 illustrates the plant operation overview. The shutdown conditions

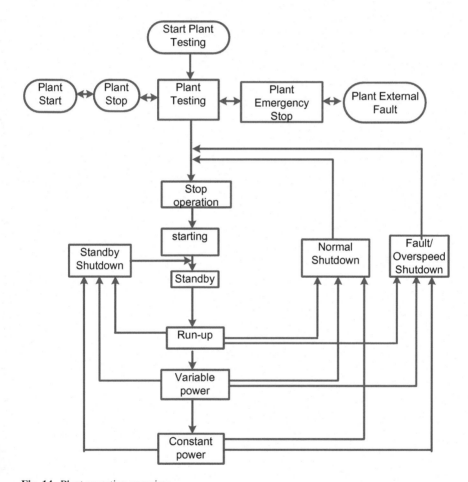

Fig. 14 Plant operation overview

of this operating state are constantly checked, and the necessary messages appear in the continuous power operation. If the rotor speed is less than the minimum pre-set value by the outer ring control system to maintain continuous power operation, the single wind turbine generator returns to the variable power operating state.

10.8   Standby Shutdown

The variable power and the continuous power operation should be possible at all times to start the standby mode of the individual wind turbines and bring it to the operating state in the standby mode and report the appropriate messages. If the average available wind speed is less than the minimum value previously determined by the outer ring control system, the wind turbine will start. Standby mode and individual wind turbines are turned off if "standby" is on. Failure stop/excessive speed, torque, or speed conditions are continuously checked in a repeated sequence. After turning off the standby mode successfully, the individual wind turbines return to "standby" state.

10.9   WPP Normal Shutdown

The plant shall be capable of discontinuation of any operating condition. The normal shutdown process is similar to the shutdown process except for the generator contact switch being opened later during the normal shutdown process. If the speed has fallen below the minimum value previously determined by the external loop control system (<2.5%), then the wind turbine brake will be mechanical brake and the wind turbine is off. Even during a shutdown state, "overspeed failure/stop" and braking conditions should be checked and status messages displayed.

10.10   Over-Speed/Fault Shutdown

Fault conditions influence the design and lifetime of wind power plant due to their transient effects. WPP is shutdown in order to prevent the components from being damaged or destroyed. Safe shutdown of the wind turbine takes place in the event of any electrical problems. Fault Shutdown procedure is similar manner to Normal Shutdown.

The over-speed shutdown can be started in case of higher operating speeds that violate the upper speed limit. Therefore, the mechanical braking system must be wisely taken into consideration in its power rating and duty-cycle rating factors.

10.11 Overspeed/Over-Temperature

When the wind power plant is in "Constant-Power" operation, i.e. at wind speeds above the nominal range, the speed is kept within the permissible range by extended range of speed control. If the operating speed still exceeds above the maximum upper limit speed (e.g. higher than 150% over nominal value), a "Fault/Overspeed Shutdown" is started.

In case of over temperature, WPP components are designed such that in normal operation, no critical high temperatures occur. If the temperature violates limits, then it is realized that there is a fault or overload in the system. Therefore, the "Normal Shutdown" operation must be initiated and the proper message is displayed.

11 Condition Monitoring for Wind Farms

Wind turbines are often subject to intense mechanical stress. CMS system ensures stability, long service life, and optimal design of wind turbine components (rotor blades, drive assemblies, inverters …). Thus, it prevents complete failure, is costly, and allows significant savings. Failure of the gearbox accounts for a large portion of the wind turbine's downtime. Therefore, gaining reliability and efficiency is the important key. Thus, measurements can be used as a tool to increase both. With monitoring, it is possible to track component status and detect potential malfunctions in a timely manner, thus preventing damage and possibly increasing the service life of the components.

11.1 Fault Diagnosis and Prognosis

A defect is a physical defect, or defect that occurs within the system. This may cause a failure: Some due or expected actions are not performed. Error detection is the determination of errors in the system and the time of detection. Error isolation is the determination of the type, location, and time of the error. Error detection follows, and includes error isolation and identification [38]. The forecast is an approach that combines information about the current state of each device with historical data from machines of the same class, physical models of failure components, and expected short-term use to predict the probability of failure of this individual device in the future. That is, the forecast gives probability expectations for each machine, which allows a strategy to balance the risks of operating the machine and indicators of damage against revenue lost while waiting for maintenance [39]. The data analysis of wind turbine using SCADA is illustrated in Fig. 15.

Data mining is a process for extracting useful information and patterns from big data. It is also termed as knowledge discovery process, knowledge mining from

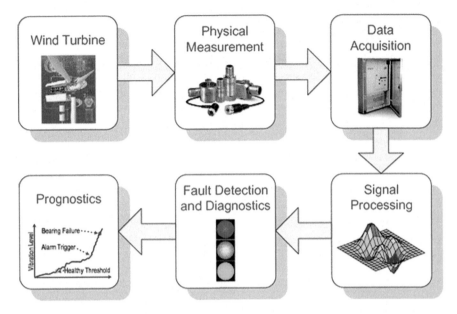

Fig. 15 Wind turbine data analysis using SCADA data

data, knowledge extraction, or data/pattern analysis [40]. The crucial goal in fault diagnosis and diagnosis is to determine the suitable maintenance approach.

11.2 Fault Diagnosis and Prognosis Systems on WT

Condition monitoring (CM) can detect errors early and prevent major malfunctions. This is associated with a significant decrease in maintenance costs. Moreover, it allows improving maintenance schedules, thus reducing downtime and enhancing equipment availability, safety, and reliability [41]. An attention on the technique of sensing (data acquisition and analysis to diagnose errors was focused [42]. Important vital information in the field of CM was provided [43]. Major malfunctions of the WT are generated due to the main gearbox; generator, main bearings and rotor blades and the possibility of malfunctions in terms of proportions are 32%, 23%, and 11% and less than 10% respectively as defined by the insurance company German Lloyd. CMS is a tool that provides component status information and can also predict expected failure/error. Table 1 summarizes the diagnostic malfunctions used and the techniques used in CMS on different parts of the WT. Digital filtering, modeling, signal, and spectrum analysis are key parts of data processing in CMS [39–42, 44]. The next step is to predict the component life and to adopt a suitable maintenance strategy.

Table 1 Summary of CMS on WT [38]

Part of WT	Fault	Technique	Sensor/monitoring quantity
Gear box	Gear tooth damages, bearing faults	Vibration monitoring and spectrum analysis, AE sensing detects pitting, cracking	Transducers, velocity sensors, accelerometers, spectral emitted energy sensors
		Oil analysis	Temperature, moisture, contamination
Generator	Stator, bearing, rotor inside	Current signature analysis	Current measurement
Rotor blades	Creep and corrosion, imbalance, fatigue, roughness	Radiography, Shearography, AE sensing	AE sensors, strain gauges, Fiber bragg grating
Tower and blades	Ultrasonic testing techniques	Time–frequency techniques and wavelet transforms	Ultrasonic sensors, Fiber brag grating
Pitch mechanism, Yaw system, power electronics/electrical system	Current and voltage analysis, electrical resistance	Spectrum analysis, eddy current, thermography	Current and voltage measuring equipment

11.3   SCADA Based Condition Monitoring of WT

Condition monitoring systems essentially offer the necessary sensor and capability of data capture required for monitoring. These systems enable diagnostic and fault detection algorithms to be installed at the sensor or RTU mounted on the turbine. Thus the gathered SCADA data have to be analyzed in order to realize the overall health of the wind turbine as well as its related components. An operational wind farm typically generates vast quantities of data. The SCADA data contain information about every aspect of a WPP including output power and wind speed and any other error registered within the system. SCADA data effectively provide early warnings of possible failures Fig. 16 shows the fault diagnosis framework. Typical parameters recorded by SCADA on a WT could be broadly categorized into wind parameters, performance parameters, vibration parameters, and temperature parameters. These parameters could be used in fault diagnosis and prognosis activity. The wind parameters are wind speed and wind deviations. The performance parameters include output power, rotor speed, and blade pitch angle. The vibration parameters comprise tower acceleration and drive train acceleration. The temperature parameters include bearing temperature and gearbox temperature. Measurement SCADA data, vibration monitoring could be used for CM [43]. One can use the combination of abnormal detection and data-trending techniques summarized in a multi-agent framework for the improvement of a fault detection scheme for WTs.

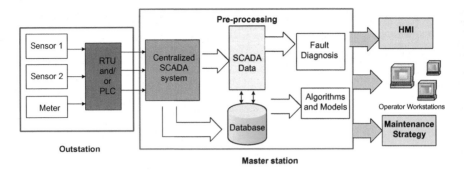

Fig. 16 Intelligent frame work for fault diagnosis and prognosis of WTs

11.4 AI Methods for Analysis of SCADA Data from WTs

Pre-processing of SCADA is a must for extraction of useful information and patterns from huge data. The various AI methods being used for analysis of SCADA data from WTs are artificial neural networks (ANNs), fuzzy systems, and arrangement techniques like adaptive neuro fuzzy inference systems (ANFIS). Figure 17 summarizes different methods.

ANNs can be used for a wide range of applications. They are inspired by the mechanism of the brain and can be classified by diverse categories that depend on the learning mechanism. Some of the key characteristics for neural networks (NNs) are their high processing speeds which are due to their massive parallelism, their demonstrated ability to be trained and produce instantaneous and correct responses from noisy or partially incomplete data, and their ability to generalize information

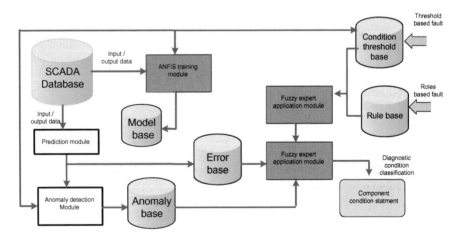

Fig. 17 Block diagram of AI methods in WT SCADA

over a wide range. These features make them a good choice for applying to WTs' data analysis.

Fuzzy systems are very useful in two general contexts like in situations involving highly complex systems whose behaviors are not well understood and in situations where an approximate, but fast solution is desired. A further advantage of fuzzy systems is that the existing expert knowledge can be implemented to improve the approximation by tuning, removing, or adding of membership functions and rules.

Fuzzy neural networks have shown to be very advantageous in dealing with real-world problems. These neuro-fuzzy systems combine the benefits of these two powerful paradigms into a single capsule. This gives the capability to accommodate both data and present expert knowledge about the issue under investigation. Recently, ANFIS has suggested for WT condition monitoring. For this purpose, ANFIS normal behavior models for common SCADA data were developed in order to detect abnormal behavior of the gathered signals and specify component faults or malfunction using the error prediction.

12 SCADA Based Abnormal Detection of Wind Turbine

The SCADA system changes the operating mode of wind farm systems with a healthy work environment and reduces operating and maintenance costs. However, a wide range of high dimensions and many types of data are not fully used or developed; only staying on data in real-time and statistics reporting historical data are usually monitored or collected. Therefore, it is important to make full use of the data collected by the electrical and electronic control systems were identified as most likely to fail, but gearbox and generator malfunctions caused the longest downtime. Figure 18 shows the components of condition monitoring system of wind turbine.

Several researchers have conducted research on observing large wind turbines and diagnosing malfunctions [5], based on a statistical learning method for detecting abnormal situations through a weighted least-square wind turbine response model to

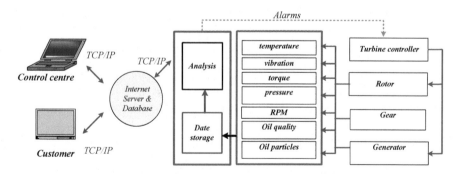

Fig. 18 Condition monitoring system for wind plant

support a vector-based wind power generator and external slope conditions [6]. The results showed that the model is better than traditional prediction methods.

Based on statistical analysis, it generally requires large data sets to provide meaningful indicators: Therefore, the most common view is that SCADA can detect initial errors at a late stage [39] using artificial neural networks, for their ability to reconstruct nonlinear dependency between Input and output, and simple formulas for diagnosing faults that occur at the gearbox level. Datasets used have a 10-min sampling time for common SCADA control systems; Gearbox vibrations and gearbox temperatures are defined as the model's target output. The time accuracy of SCADA will turn out to be very rough for reliable vibration analysis, which should be observed somewhat on its appropriate time scale (several Hz). At present, data mining methods such as agglomeration and statistical model are widely used in domestic and foreign companies, but the cleaning process is complicated and the cleaning conditions are harsh [45–49]. Therefore, in order to perform a reliable analysis of the power generation performance of wind turbines, an effective and varied cleaning method is urgently needed. In light of this, this chapter first extracts the features from the big data and high dimensions that SCADA collects and removes the unrelated and unnecessary parameters.

13 Data Mining of Characteristic Parameters for Wind Turbines

The data collected and recorded by SCADA wind turbine system has high dimensional properties. Figure 19 illustrates obtaining typical data for wind turbines. Therefore, in this chapter, a data mining algorithm based on the degree of gray correlation [46] has been proposed to overcome the above-mentioned shortcomings and to improve the accuracy and effectiveness of wind turbine operating condition assessments.

Variables in wind turbine information are recorded by the SCADA system. Figure 18 shows the observing variables collected by the SCADA wind turbine system and its response code. The aim of the study in this chapter is that wind turbines are in an unlimited state of energy and healthy running. They have some monitoring quantities such as the control and alarm mode for some parameters recorded in SCADA system, speed mode, column 1 status, column 2, axis 3, etc. Variables can be ignored in a fixed state. Table 1 is part of the parameter alarm information from the GE wind turbine manufacturer. To address these self-vectors beforehand, we must remove these self-vectors to avoid the dimensional disaster caused by many features.

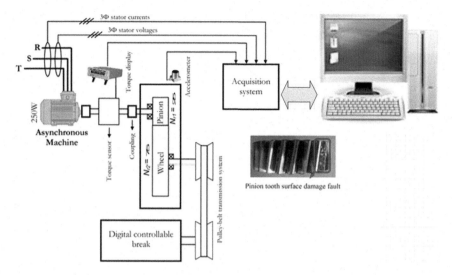

Fig. 19 Data acquisition in wind turbine

13.1 Communication Network for Wind Power Farms (WPPs)

The SCADA system enables operators to monitor, control, and record wind power plant data from a remote location called a central control station [1, 2]. It consists of three main components as shown in Fig. 20. Inside the turbine tower, including wind turbine controller (WTC), remote terminal units (RTU), smart electronic devices (IEDs), and sensors. The World Trade Center collects all data using short communication links, making them available for processing and transfers to the control center. A closed circuit television system (CCTV) and internet connection can share or use a separate network between the wind turbine and the control center.

A communications network is based on Ethernet equipment (Gigabit Ethernet), to transfer data between teams and the control center. Most WPP s use the same SCADA power cable paths as they used for power distribution. The Control Center connects individual work teams and meteorological stations with the Control Center. Operators manage the behavior of all teams as a whole. It requires a long distance to transfer data.

13.2 Ethernet Passive Optical Network (EPON)-Based Communication Network for WPPs

EPON consists of a control center OLT, multiple ONUs, and POS. Downstream, EPON is a point-to-multipoint network; OLT broadcast controls messages and data packets on all operating units via the passive partition unit. On the upside, EPON is a

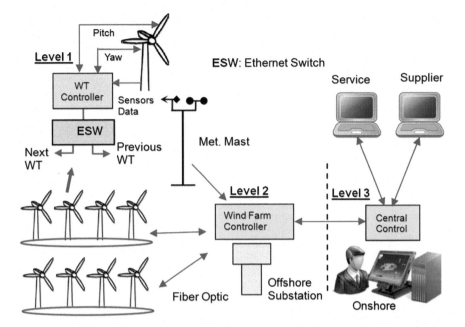

Fig. 20 Schematic view of conventional wind power farms (WPP) network

multi-point network. Multiple ONUs send data to OLT by passive combination [18]. Figure 21 shows a schematic view of the EPP-based communication network for WPP [24]. The proposed network consists of an optical network device (ONU) that is deployed on the WT side to collect various data, including wind turbine operation, meteorological data, and fault and safety parameters from different internal networks. All ONUs from different WTs connected to OLT unit, located on the side of the control center. The path between WTs and the Control Center does not contain any active elements, which saves costs and reduces the complexity of maintenance and deployment, compared to the current switched Gigabit Ethernet.

In Fig. 21, the shown architectures (star, cascade, etc.) represent the design of a new WPPs scheme. Mixed configuration (Ethernet switch and ONU) represents the modification of existing WPPs to support EPON technology, which is outside the scope of this work. In the astral configuration, four wind turbines with fibers distributed are connected to the control center using a single point of sale (1 × 4). In succession configuration, four points of sale (1 × 2) are connected in succession. One port is connected to the next weight, while the other port is connected to the WT-ONU.

For example, EPON uses a 1Gbps single fiber, with a transmission range of 20 km. Different wavelengths are used to support current flow from current to direction, 1,490 nm and 1310 nm, respectively. Each WT sends data in its own time slots, to avoid data collisions. EPON must use a media access control (MAC) mechanism, to control access to the shared media, to prevent collision of different ONU data in

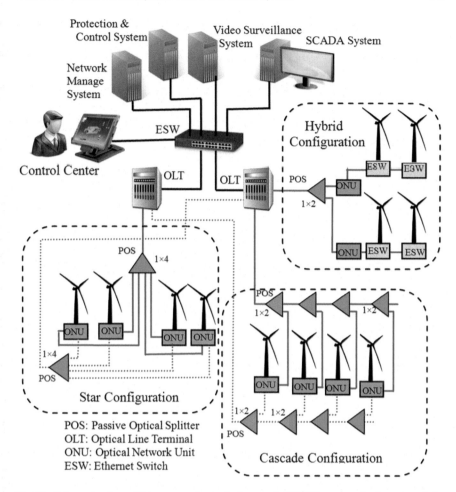

Fig. 21 Schematic view of Ethernet passive optical network (EPON) communication network for WPP

the opposite direction. OLT efficiently shares transmission bandwidth between all turbines ONUs. Failure to reach this shared channel in a timely manner can negatively affect communication and should be considered an aspect of reliable communication. There have been many studies and frameworks for managing medium access control at [1, 5, 18, 24].

14 Future Challenges

From the literature reviewed, researchers have successfully demonstrated that by tracking wind speed and energy output parameters, the overall health of the turbine can be supervised. Moreover, SCADA data can be used for CM from WTs and fixing errors. There was success in using SCADA data for power forecasting, optimum control settings, performance appraisal, turbine malfunction prediction (steering acceleration/tower acceleration/gearbox failure) as well as vibrations on WT. Many AI technologies have been applied including NN, Fuzzy, ANFIS, GA's, etc. It is also proposed to use data consolidation techniques to monitor the health of WTs' [47]. Some challenges that must be overcome before SCADA data analysis becomes completely successful: SCADA data can vary from turbine to turbine and SCADA data change with operational conditions. Therefore, it becomes difficult to distinguish between real error and false error as a big challenge.

Moreover, WT SCADA data is usually an average of 10 min data, so some information is lost. Thus with reference to the WT status monitoring research, it is suggested that a framework that takes SCADA data as well as high-frequency data from sensors (some of them) be proposed to diagnose and forecast traditional data and SCADA data. After comparing the two results, the appropriate method for making maintenance decisions can be chosen. The performance of AI-based data mining algorithms and CM algorithms are showing very promising results. Hence using mathematical intelligence models more efficient models can be obtained thereby enhancing the model's accuracy and durability. Data mining (AI-based) and evolutionary accounts are combined to build prediction and monitoring models.

Acknowledgments The editors of this book would like to thank the authors and reviewers for their contributions and efforts. Moreover, we would like to thank all colleagues from K. A. CARE Energy Research and Innovation Center, Riyadh, Saudi Arabia for their help and efforts.

References

1. Sayed K, Gabbar HA (2018) Building energy management systems (BEMS). In: Energy conservation in residential, commercial, and industrial facilities, pp 15–81
2. Sayed K, Gabbar HA (2017) SCADA and smart energy grid control automation. In: Smart energy grid engineering, pp 481–514
3. Ahmed MA, Eltamaly AM, Alotaibi MA, Alolah AI, Kim Y-C (2020) Wireless network architecture for cyber physical wind energy system. IEEE Access 8:40180–40197
4. Eltamaly AM, Al-Saud MS, Abo-Khalil AG (2020) Dynamic control of a DFIG wind power generation system to mitigate unbalanced grid voltage. IEEE Access 8:39091–39103
5. Tautz-Weinert J, Watson S (2017) Using SCADA data for wind turbine condition monitoring—a review. IET Renew Power Gener 11(4):382–394. https://doi.org/10.1049/iet-rpg.2016.0248
6. Sciacca SC, Block WR (1995) Advanced SCADA concepts. IEEE Comput Appl Power 8(1):23–28
7. Chan E-K, Ebenhoh H (1992) The implementation and evolution of a SCADA system for a large distribution network. IEEE Trans Power Syst 7(1):320–326
8. Leonardi A, Mathioudakis K, Wiesmaier A, Zeiger F (2014) Towards the smart grid: substation automation architecture and technologies. In: Advances in electrical engineering, 13 p. https://doi.org/10.1155/2014/896296
9. Hansen AD, Sørensen P, Iov F et al (2006) Centralised power control of wind farm with doubly fed induction generators. Renew Energy 31(7):935–951

10. Xie Y, Liu C, Wu Q, Li K, Zhou Q, Yin M (2017) Optimized dispatch of wind farms with power control capability for power system restoration. J Mod Power Syst Clean Energy 5(6):908–916. https://doi.org/10.1007/s40565-017-0341-9

11. Abo-Khalil AG, Alghamdi A, Tlili I, Eltamaly AM (2019) Current controller design for DFIG-based wind turbines using state feedback control. IET Renew Power Generat 13(11):1938–1948

12. Singh N, Kliokys E, Feldmann H, Kiissel R, Chrustowski R, Jaborowicz C (1988) Power system modelling and analysis in a mixed energy management and distribution management system. IEEE Trans Power Syst 13(3)

13. Noske S, Falkowski D, Swat K, Boboli T (2017) UPGRID project: the management and control of LV network. In: 24th international conference & exhibition on electricity distribution (CIRED), 12–15 June 2017, pp 1520–1522

14. Etherden N, Johansson AK, Ysberg U, Kvamme K, Pampliega D, Dryden C (2017) Enhanced LV supervision by combining data from meters, secondary substation measurements and medium voltage supervisory control and data acquisition. In: 24th international conference & exhibition on electricity distribution (CIRED), 12–15 June 2017, pp 1089–1093

15. Sayed K, Gabbar HA (2017) Supervisory control of a resilient DC microgrid for commercial buildings. Int J Process Syst Eng 4(2–3):99–118

16. Eltamaly AM, Mohamed YS, El-Sayed A-HM, Elghaffar ANA (2019) Analyzing of wind distributed generation configuration in active distribution network. In: 2019 8th international conference on modeling simulation and applied optimization (ICMSAO). IEEE, pp 1–5

17. Dhiman HS, Deb D, Muresan V, Balas VE (2019) Wake management in wind farms: an adaptive control approach. Energies 12:1247. https://doi.org/10.3390/en12071247

18. Ahmed MA, Pan J-K, Song M, Kim Y-C (2016) Communication network architectures based on ethernet passive optical network for offshore wind power farms. Appl Sci 6:81. https://doi.org/10.3390/app6030081

19. Stancu DC, Federenciuc D, Golovanov N, Stanescu D (2017) New functionalities of smart grid-enabled networks. In: 24th international conference & exhibition on electricity distribution (CIRED), Open Access Proc J 2017, no 1, pp 1903–1906

20. Huang M, Wei Z, Sun G, Zang H (2019) Hybrid state estimation for distribution systems with AMI and SCADA measurements. IEEE Access 7:120350–120359

21. Zhao H, Ma L, Yan X, Zhao Y (2019) Historical multi-station SCADA data compression of distribution management system based on tensor tucker decomposition. IEEE Access 7:124390–124396

22. Khaled U, Eltamaly AM, Beroual A (2017) Optimal power flow using particle swarm optimization of renewable hybrid distributed generation. Energies 10(7):1013

23. Vera YEG, Dufo-López R, Bernal-Agustín JL (2019) Energy management in microgrids with renewable energy sources: a literature review. Appl Sci 9:3854. https://doi.org/10.3390/app9183854

24. Ahmed MA, Kang YC, Kim Y-C (2015) Modeling and simulation of ICT network architecture for cyber-physical wind energy system. In: 2015 IEEE international conference on smart energy grid engineering (SEGE), Oshawa, Canada

25. Goraj M, Epassa Y, Midence R, Meadows D (2010) Designing and deploying ethernet networks for offshore wind power applications—a case study. In: 10th IET international conference on developments in power system protection (DPSP 2010). Managing the change, p 84

26. Li P, Song Y, Li D, Cai W, Zhang K (2015) Control and monitoring for grid-friendly wind turbines: research overview and suggested approach. IEEE Trans Power Electron 30(4):1979–1986

27. Yang JM, Cheng KWE, Wu J, Dong P, Wang B (2004) The study of the energy management system based-on fuzzy control for distributed hybrid wind-solar power system. In: Proceedings of first international conference on power electronics systems and applications, pp 113–117

28. Eltamaly AM, Farh HM (2015) Smart maximum power extraction for wind energy systems. In: 2015 IEEE international conference on smart energy grid engineering (SEGE). IEEE, pp 1–6

29. Abo-Khalil AG, Alghamdi AS, Eltamaly AM, Al-Saud MS, Praveen RP, Sayed K (2019) Design of state feedback current controller for fast synchronization of DFIG in wind power generation systems. Energies 12(12), 2427
30. Abo-Khalil AG, Alyami S, Sayed K, Alhejji A (2019) Dynamic modeling of wind turbines based on estimated wind speed under turbulent conditions. Energies 12(10):2019
31. Sayed K, Abdel-Salam M (2017) Dynamic performance of wind turbine conversion system using PMSG-based wind simulator. Electri Eng J 99:431–439
32. Sayed K, Gabbar H (2016) Smart distribution system Volt/VAR control using the intelligence of smart transformer. In: Proceedings of the 4th IEEE international conference on smart energy grid engineering SEGE 2016, pp 52–56
33. Abdel-Salam M, Ahmed A, Ziedan H, Sayed K, Amery M, Swify M (2011) A solar-wind hybrid power system for irrigation in Toshka area. In: IEEE Jordan conference on applied electrical engineering and computing technologies AEECT, Amman, Jordan, pp 38–43
34. Sayed K, Kassem AM, Aboelhassan I, Aly AM, Abo-Khalil AG (2019) Role of supercapacitor energy storage in DC microgrid. In: 1ST international conference on electronic engineering Iceem 2019, Egypt, 7–8 December 2019
35. Sayed K, Kassem AM, Aboelhassan I, Aly AM (2019) Energy management and control strategy of DC microgrid including multiple energy storage systems. In: 21st international Middle East power systems conference (MEPCON), Tanta University, Egypt
36. Praveen RP, Therattil J, Jose J, Abo-Khalil A, Alghamdi A, Bindu GR, Sayed K (2020) Hybrid control of a multi area multi machine power system with FACTS devices using non-linear modelling. IET Generat Trans Distrib 14(10), 1993–2003
37. Sayed K, Abo-Khalil AG, Alghamdi AS (2019) Optimum resilient operation and control DC microgrid based electric vehicles charging station powered by renewable energy sources. Energies 12:4240. https://doi.org/10.3390/en12224240
38. Wang K-S, Sharma VS, Zhang Z-Y (2014) SCADA data based condition monitoring of wind turbines. Adv Manuf 2:61–69. https://doi.org/10.1007/s40436-014-0067-0
39. Kusiak A, Li W (2011) The prediction and diagnosis of wind turbine faults. Renew Energy 36(16–23):30
40. Eltamaly AM, Alolah AI, Farh HM, Arman H (2013) Maximum power extraction from utility-interfaced wind turbines. New Develop Renew Energy 159–192
41. Zhang Z, Kusiak A (2012) Monitoring wind turbine vibration based on SCADA data. J Sol Energy Eng 134(021004):32
42. Elnozahy A, Sayed K, Bahyeldin M (2019) Artificial neural network based fault classification and location for transmission lines. In: 2019 IEEE conference on power electronics and renewable energy (CPERE), Aswan City, Egypt, pp 140–144. https://doi.org/10.1109/CPERE45374.2019.8980173
43. Dempsey PJ, Sheng S (2013) Investigation of data fusion applied to health monitoring of wind turbine drivetrain components. Wind Energy 16(4):479–489
44. Eltamaly AM, Khan AA (2011) Investigation of DC link capacitor failures in DFIG based wind energy conversion system. Trends Electri Eng 1(1):12–21
45. Wilkinson M, Darnell B, Harman K (2013) Presented at EWEA 2013 annual comparison of methods for wind turbine condition monitoring with SCADA data. EWEA 2013 annual event, Vienna, pp 4–7
46. Schlechtingen M, Ferreira Santos I (2011) Comparative analysis of neural network and regression based condition monitoring approaches for wind turbine fault detection. Mech Syst Signal Process 25:1849–1875
47. Pandit R, Infield D (2018) SCADA—based wind turbine anomaly detection using Gaussian process models for wind turbine condition monitoring purposes. IET Renew Power Generat 12(11). https://doi.org/10.1049/iet-rpg.2018.0156
48. Mokryani G, Siano P, Piccolo A, Cecati C (2011) A novel fuzzy system for wind turbines reactive power control conference paper. In: Proceedings of IEEE international conference on fuzzy systems, Taipei, Taiwan, 27–30 June 2011

49. Laina R, El-Amrani A, Boumhidi I (2019) Finite frequency H ∞ control for T-S fuzzy systems: application to wind turbine. In: Third international conference on intelligent computing in data sciences (ICDS), Oct 2019
50. Almutairi A, Abo-Khalil AG, Sayed K, Albagami N (2020) MPPT for a PV grid-connected system to improve efficiency under partial shading conditions. Sustainability 12(24):10310

Harmonic Source Detection for an Industrial Mining Network with Hybrid Wind and Solar Energy Systems

Rosalia Sinvula, Khaled M. Abo-Al-Ez, and Mohamed T. Kahn

Abstract In modern electrical networks, grid connection of renewable energy systems brings more challenges, the most related to power systems operators and controllers are the power quality (PQ) issues. In developing countries, solar photovoltaic (PV) systems are widely used as compared to wind energy systems due to certain legislative and economic constraints. Therefore, wind energy PQ impacts are less analyzed compared to solar PV ones. But with the expected increase of wind energy penetration at the distribution networks in these countries such as the Southern Africa countries, it is inevitable to pay attention to wind energy PQ effects. On top of the PQ issues, the issue of harmonic distortion is a major concern especially with the use of power electronics interface of renewable energy systems, and the wide use of non-linear loads. The quality of power for the distribution network is minimized because of harmonic distortion. The total harmonic distortion (THD) of the current and voltage is exceeding the harmonic limits specified by the national and international standards. The harmonic sources in the distribution network need to be found whether it is from the upstream or downstream of the network. The dominant harmonic order needs to be found as well, as this plays a major factor in designing the harmonic mitigation for the distribution network. The comprehensive wind turbine modeling in the DIgSILENT software package is presented in this chapter, along with a novel model for Variable Speed Drives (VSDs) which is used for harmonic analysis. THD at the Point of Common Coupling (PCC) is analyzed when the numbers of wind farms are increasing, using a typical case study of a network with mining industries

R. Sinvula · K. M. Abo-Al-Ez (✉) · M. T. Kahn
Faculty of Engineering and the Built Environment, Centre for Power Systems Research (CPSR), Cape Peninsula University of Technology (CPUT), Symphony Way, Bellville Campus, Cape Town 7535, South Africa
e-mail: aboalezk@cput.ac.za

R. Sinvula
e-mail: 207080704@mycput.ac.za

M. T. Kahn
e-mail: khant@cput.ac.za

© The Author(s), under exclusive license to Springer Nature Switzerland AG 2021
A. M. Eltamaly et al. (eds.), *Control and Operation of Grid-Connected Wind Energy Systems*, Green Energy and Technology,
https://doi.org/10.1007/978-3-030-64336-2_7

153

in Southern Africa. Lastly, this chapter proposes modified harmonic source detection based on the direction of active power flow which is practical and commercially viable.

Keywords Power quality · Harmonics · Source detection · Wind · Solar · Mining · Variable speed drives · DIgSILENT modeling

1 Overview

The current high penetration of renewable energy sources (RES) into the power systems is causing a significant impact on the Quality of Supply (QoS) [1, 2]. Hence, engineers and system operators are facing new and evolving challenges concerning power system reliability, quality, and security. Wind energy is one of the promising RES that is widely employed due to its technical potentials and economic viability [3–5]. Various stakeholders must understand the wind power plants power quality (PQ). These stakeholders are; the manufacturers of the wind power plant components, as they need to ensure that their products comply with the PQ standards, the purchasers have to include the correct specifications during the tendering process, the operators, planners, and regulators need to verify the requirements specified by the purchaser. Among different PQ challenges, harmonic pollution caused by grid-connected wind generation systems is the main concern to both power utilities and customers. Wind power plants have low harmonic emission levels at the higher harmonic order [6, 7]. The harmonics that exceed the harmonic limits are the low-order harmonics such as the 5th, 7th, 11th, and 13th [3, 7]. The relationship between the wind farm harmonic emission and its fluctuating output power depends on a correlation index which ranges between 0.7 and 1.0 [3, 7].

On the load side, non-linear loads such as the variable speed drives (VSDs), electric arc furnaces, saturation of magnetization of transformers, etc., are the main cause of the harmonic distortion [8–10]. The harmonics rise when the current generated by these loads changes based on the impedance of the power system [11]. Harmonics have negative impacts on the power system such as the increase of power losses, deterioration of the insulation, and reduction of the life expectancy of the equipment [12–15]. Thus, the communication equipment starts to malfunction because of harmonic distortion. There are national and international standards that give the harmonic limits of the current and voltage [16–18]. Different harmonic filters are implemented in the distribution network to limit the harmonics level at the Point of Common Coupling (PCC). When the wind energy conversion system, composed mainly of the turbine and the generator, is connected to the power grid using power electronic converters, the harmonic analysis is needed to determine the harmonic distortion of the current and voltage, especially when wind farms are in proximity to the loads [19]. When the wind energy system is in proximity to the customers, connected at the distribution network, the PQ impacts are higher than when it is connected at the transmission network of the high voltage(HV) and extra-high voltage (EHV) system. The distance between the wind energy systems and PCC has effects on the customer's PQ as it is normal up to a few kilometers. The connection of the wind energy system on HVor medium voltage (MV) transmission lines has economic constraints but has fewer effects on the customer's load PQ [11]. According to [11], the wind energy system shall consist of voltage and frequency relays to disconnect it when abnormal values of voltage and frequency are noticed. However, this does not specify the definition of abnormal frequency, and whether it includes the harmonic frequency as part of the abnormal frequency. If the harmonic frequency is part of the abnormal frequency, then the wind energy system will always be disconnected as there is no network without the harmonic frequency. This frequency relay must be set for the harmonic limits based on the national and international standards.

This chapter's emphasis is on harmonic distortion, and how to detect the source of harmonics when there is higher penetration of wind energy as part of the power grid. The specifications of wind energy systems that can be connected at the grid are part of the electricity grid code of each country. These should be used during the planning, maintenance, and operation of the power grid. This chapter is organized as follows:

In Sect. 2, harmonic distortion concepts are introduced; in Sect. 3, a modified active power flow based harmonic source detection method is presented; Sect. 4 presents the modeling of an industrial network case study using DIgSILENT; Sect. 5 introduces the detailed simulation results and discussions of the different case studies applied to validate the proposed concepts and algorithms; lastly, Sect. 6 presents the overall conclusion of the chapter.

2 Harmonics Distortion

As smarter power systems are being implemented, the investigation of harmonic challenges becomes a priority PQ issue to be investigated. Figure 1 illustrates the current waveforms of the individual harmonic frequency and the combined frequency (hch) (fundamental and harmonic frequencies) [20]. The linear load (non-distorted load) does not cause harmonic distortion although equipment can be affected by the harmonic injected by the distorted load (non-linear load) connected at the same PCC or anywhere along with the network [21].

A voltage drop for each harmonic order is caused by the harmonic current passing through an impedance which results in voltage harmonics occurring at the PCC bus. This voltage distortion is controlled by the system impedance and the current through the load equipment. The load has no control of the voltage distortion that is caused by the load current harmonics. Fourier analysis is used to analyze the harmonic components of the distorted current and voltage waveforms under steady-state conditions [9]. The distorted waveforms of positive and negative half-cycles are identical. Only odd harmonic is used in Fourier analysis, which is called combined frequency or characteristic harmonics (hch). Most of the harmonic distribution sources only generate odd harmonics; if even harmonics are noticed, it shows the error with the system if

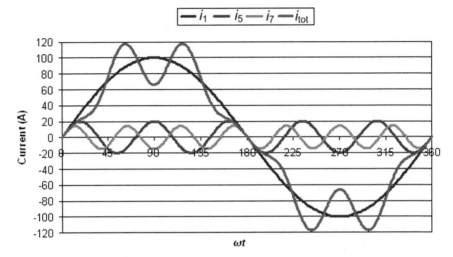

Fig. 1 The current waveform of the combined frequency (hch) [20]

the supply voltage is symmetrical. In a modern power system, it is unlikely to find the ideal supply voltage consisting of some percent of background harmonic from the rest of the network. The magnitude of the harmonic current component (I_h) is inversely proportional to and given expression by [9]:

$$I_h = \frac{I_1}{h} \tag{1}$$

where: I_1 is the current magnitude at the fundamental frequency (f_1); h is the harmonic number and is equivalent to hch ($h = hch$); however, when only characteristic harmonics are present, f_1 is a first harmonic then $f_1 = h_1$. It makes the explanation easier to use h_1 instead of f_1. The characteristic harmonic current of a three-phase power system is calculated using the expressions respectively as per Red-White-Blue (R-W-B) [22]:

$$
\begin{aligned}
i_R = {} & I_{1m}\cos(\omega t) - I_{5m}\cos(5\omega t) + I_{7m}\cos(7\omega t) \\
& - I_{11m}\cos(11\omega t) + I_{13m}\cos(13\omega t) \\
& - I_{17m}\cos(17\omega t) + I_{19m}\cos(19\omega t) \ldots
\end{aligned} \tag{2}
$$

$$
\begin{aligned}
i_W = {} & I_{1m}\cos(\omega t - 120°) - I_{5m}\cos(5\omega t - 240°) + I_{7m}\cos(7\omega t - 120°) \\
& - I_{11m}\cos(11\omega t - 240°) + I_{13m}\cos(13\omega t - 120°) \\
& - I_{17m}\cos(17\omega t - 240°) + I_{19m}\cos(19\omega t - 120°) \ldots
\end{aligned} \tag{3}
$$

$$
i_B = I_{1m}\cos(\omega t + 120°) - I_{5m}\cos(5\omega t + 240°) + I_{7m}\cos(7\omega t + 120°)
$$

$$- I_{11m} \cos(11\omega t + 240°) + I_{13m} \cos(13\omega t + 120°)$$
$$- I_{17m} \cos(17\omega t + 240°) + I_{19m} \cos(19\omega t + 120°) \ldots \tag{4}$$

Equations (2), (3), and (4) are used to determine the waveform of the complex current for three-phase systems. The complex current waveform illustrated in Fig. 1 is like the distorted current waveform for the three-phase six-pulse rectifier/inverters that have the 5th and 7th harmonic orders as dominant.

The harmonic indices need to be scrutinized when it comes to harmonic analysis, as this gives enhanced, well-focused information about the components involved in the calculation for THD. The THD calculated and analyzed can be either for current or voltage [9].

$$\%THD = \frac{\sqrt{\sum_{h>1}^{h_{max}} F_h^2}}{F_1} \times 100\% \tag{5}$$

where: F_h is a harmonic component root mean square (rms) value, F can be voltage or current and F_1 is the voltage or current magnitude at $h = 1$. Individual harmonic component of current or voltage magnitude is expressed as [9]:

$$\%HD = \frac{F_h}{F_1} \times 100\% \tag{6}$$

Equation (5) and (6) can be expressed in terms of current and voltage [9];

$$\%I_{THD} = \frac{\sqrt{\sum_{h=2} I_h^2}}{I_1} \times 100\% \tag{7}$$

$$\%V_{THD} = \frac{\sqrt{\sum_{h=2} V_h^2}}{V_1} \times 100\% \tag{8}$$

$$\%HD = \frac{I_h}{I_1} \times 100\% \tag{9}$$

$$\%HD = \frac{V_h}{V_1} \times 100\% \tag{10}$$

These equations are used in most power system software packages to calculate the individual harmonics and THD of current or voltage, based on the fundamental frequency current or voltage magnitude. The harmonic distortion in waveforms can be caused by the combined frequency which consists of fundamental and harmonic frequencies. The causes of harmonic distortion are non-linear loads and RES such as solar photovoltaic, wind, etc. Its effects are severe damage to capacitors and transformers, degradation of motor insulation, control equipment malfunctions, and decreases in performance and efficiency of the power system. Once the harmonic distortion is not limited to specified harmonic limits as specified in the standards, then the issue of harmonic source detection becomes a power systems challenges.

3 Harmonic Source Detection Method

The main target of most power utility companies in the world is to ensure the quality of power supply. There are many PQ challenges, however, one of the most concerned is the harmonic distortion that can result either from upstream (utility side) through background harmonics or downstream (customer's side) through non-linear loads or RES. Worldwide utilities and customers are more concerned with the detection of harmonic sources at the PCC. The modern power system consists of multiple customers with different operating conditions that are connected to the PCC. According to [23], most studies concentrate on the detection of the harmonic source between a customer and utility. However, this does not stand for practical power system with multiple customers with different load conditions connected to the PCC. The location of harmonic source detection has been part of research since the 1990s [24] when the power system became automated with multiple non-linear loads. There are various techniques and methods reported in the literature which were developed to detect harmonic source such as in [25–41]. These methods and techniques have been discussed and scrutinized concerning certain issues. Hence, researchers have reviewed and compared different methods and techniques as in [42–48]. Three categories of these methods have been classified as methods based on the direction of active power flow, reactive power, and voltage–current ratio. In this chapter, the focus is on active power flow method, because it is a common method used by most industries to determine the flow of harmonics in the distribution network and for data monitoring analysis. This method is integrated into most PQ analyzers and in commercial software.

3.1 The Modified Direction of Active Power Flow

It is important to examine the harmonic power flow direction method and quantify the size of the harmonic contribution of each harmonic source connected at the Point of Connection (PoC). The harmonic power flow depends on the flow of harmonic current within the power system. It can be that the harmonic sources have negative power at a harmonic frequency which means it injects the harmonic current, even though the dominant contribution might only be from one or two harmonic sources. Thus, the load with the highest percentage of individual and THD is the dominant contributor to harmonics. The sign of active power should be considered when calculating the total active power at the hch. The total active power of the non-linear loads and RES is affected by the harmonics compared to the total active power of linear loads. A non-linear load and RES draw fundamental power (P_1) and inject harmonic power (P_h);

the harmonic power can be in either direction. Similarly, for linear load, harmonic power (P_h), is not exactly at the same phase as the fundamental power (P_1), thus, the direction of active power flow is used [49]. When harmonic distortion is present in the network the power should include harmonic power with its flow of directions either positive or negative [49]. This will result in a total power (P_T) to be either one of Eqs. (11)–(12) as for non-linear loads or RES and linear load. These equations calculate the total active power at the hch [50]:

$$P_T = P_1 - \sum P_h \tag{11}$$

$$P_T = P_1 + \sum P_h \tag{12}$$

Harmonic power (P_h), is obtained by Eq. (13) in [9], where θ_h is voltage angle and $\emptyset_h$ is the current angle.

$$P_h = \sum_{h=1}^{\infty} V_h I_h \cos(\theta_h - \emptyset_h) \tag{13}$$

The harmonic order or hch that are part of the network depends on the size of the rectifiers. The rectifiers are based on the pulse number. For instance, the determination of the dominant harmonic for a 12-pulse rectifier is achieved by Eq. (14) in [9].

$$h = (n \times p) \pm 1 \rightarrow n\text{th and } m\text{th } harmonics \tag{14}$$

where n is integers (1, 2, 3, 4...), and p is the rectifier pulse number. Thus, the characteristic harmonic of the 12-pulse rectifier can be obtained using Eq. (14) as follows:

$$h = (1 \times 12) \pm 1 \rightarrow 11\text{th and } 13\text{th } harmonics$$
$$h = (2 \times 12) \pm 1 \rightarrow 23\text{th and } 25\text{th } harmonics$$
$$h = (3 \times 12) \pm 1 \rightarrow 35\text{th and } 37\text{th } harmonics$$
$$h = (4 \times 12) \pm 1 \rightarrow 47\text{th and } 9\text{th } harmonics$$
$$h = (n \times 12) \pm 1 \rightarrow n\text{th and } m\text{th } harmonics$$

The modification of the direction of active power flow must be conducted in three stages as follows,

(a) Identify the side of the harmonic source and quantify each side's contribution;
(b) Identify the contributor in case of the customer side and quantify each customer; and
(c) Prove the dominant harmonic order of the contributor.

To determine the harmonic source, the meaning of active power is summarized in Tables 1 and 2.

Table 1 Harmonic source meaning based on the total power at the PCC

Total power	Equation	Sign	Harmonic source
	$P_{T\_Pcc} > 0$	Positive	Utility
	$P_{T\_Pcc} < 0$	Negative	Customer

Table 2 Harmonic source meaning based on the individual harmonic power at the PCC from the upstream and downstream of the network

Power of individual harmonic	Equation		Sign	Harmonic source
	Upstream	$P_{Pcc} > 0$	Positive	Customer
		$P_{Pcc} < 0$	Negative	Utility
	Downstream	$P_{Pcc} > 0$	Positive	Utility
		$P_{Pcc} < 0$	Negative	Customer

Table 3 Meaning of the individual harmonic power at the utility grid or RES

Power of individual harmonic at the utility grid or RES	Equation	Sign	Generating/Absorbing
	$P_h > 0$	Positive	Generating
	$P_h < 0$	Negative	Absorbing

Table 4 Meaning of the individual harmonic power at the customer or load

Power of individual harmonic at the customer or load	Equation	Sign	Generating/Absorbing
	$P_h > 0$	Positive	Absorbing
	$P_h < 0$	Negative	Generating

The meaning of individual harmonic power for utility grid or RES and customer (load) is determined in Tables 3 and 4.

The DIgSILENT software package is used for simulation investigations and to prove the effectiveness of the modified method developed for harmonic source detection at the PCC where multiple customers are connected. It forms the network drawing functions, modeling, and features enabling power system analyzes to be conducted. This software package can show the direction of active power, and generate results of certain parameters at different frequencies. It can conduct the load flow, harmonic, and dynamic analysis as well as motor starting, to mention a few of its capabilities. Thus, it is an effective software package that most commercial industries use and it is best for harmonic source detection as it gives the individual and THD of the current and voltage at different points in the network. In the following section, detailed modeling of a case study for harmonic source detection of a typical industrial network in Southern Africa is presented using DIgSILENT.

4 Modeling of an Industrial Network Case Study Using DIgSILENT

The industrial network consists of the two mines, 5 MW solar PV plant, 5 MW wind farm composed of *fully rated converter* wind generators, and different commercial loads. It has a 400 kV slack bus with short-circuit power of Ssc = 10,000 MVA, short-circuit current of Isc = 14.43 kA, and X/R ratio of 10. The frequency dependencies for the resistance and reactance of the positive sequence components are obtained through conducting the frequency sweep or frequency scan. The two mines are Mine A (open-pit) which consists of four three-phase 12-pulse rectifiers, and Mine B (underground) which consists of many three-phase 6-pulse VSDs. The VSD model is not directly available in DIgSILENT, therefore the development and modeling of VSDsare firstly presented in the following section.

4.1 Variable Speed Drives (VSDs)

There are five industrial fans situated underground for cooling purposes, which are controlled by the VSDs that control the speed and the torque of the AC induction motors. The VSDs also reduce the starting transients for the induction motors that are used for pumps or fans. The VSD model is represented by a rectifier, DC capacitor, and pulse width modulation (PWM) connecting to a three-phase induction motor. It is very crucial to model the VSDs accurately for harmonic analysis due to the harmonic current generated by the rectifiers. The active power of the DC system is calculated by [51, 52]:

$$P_{DC} = V_{DC} \times I_{DC} \tag{15}$$

The commutation reactance (X_{cr}) is vital as it causes the DC voltage drop, which is subtracted from a nominal DC voltage [51].

$$\Delta V_{DC} = \frac{3}{\pi} X_{cr} \times I_{DC} \tag{16}$$

These equations are used in the calculations of six-pulse three-phase diode rectifiers or three-phase line commutated thyristor-controlled rectifiers [51].

$$V_{DC} = 1.35 \times V_{LL} - \Delta V_{DC} \tag{17}$$

where: V_{LL} is the line-to-line voltage. To calculate the AC voltage, the following formulae are used:

$$V_{LL} = \frac{\sqrt{3}}{2\sqrt{2}} V_{DC} = 0.612 V_{DC} \tag{18}$$

$$V_{rms} = \frac{1}{\sqrt{2}}V_{DC} = 0.707V_{DC} \qquad (19)$$

The rectifier is modeled as a current source load with an ideal six-pulse rectifier harmonic spectrum; this keeps the DC voltage constant. The VSD dynamic model is shown in Fig. 2 [53].

The graphic design of the VSD in the DIgSILENT is given in Fig. 3 [54]. The developed model of a VSD created from the frame definition **VSD_V/f_PWM** as indicated in Fig. 4. The VSD control model is shown in Fig. 5 based on [54].

To obtain the graph for the start-up process of the induction motor, the simulation for electromagnetic transients/stability analysis functions (RMS/EMT) is conducted besides the VSD speed control parameters and its equations. Although the focus of this research is on the harmonic study, it is vital to understand and check how the induction motor starts and how its speed is controlled by the VSD. This is discussed in the following section.

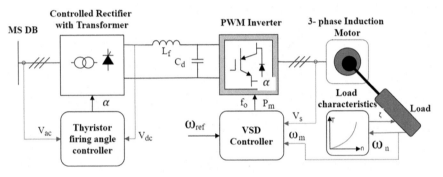

Fig. 2 VSD driven motor load dynamic simulation model [53]

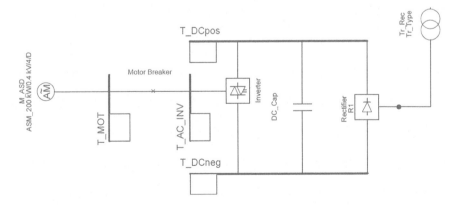

Fig. 3 Graphic design of the VSD in DIgSILENT [54]

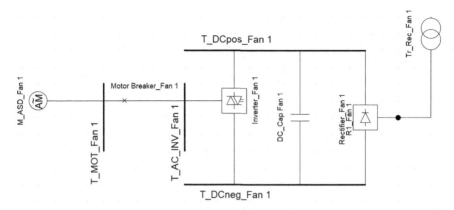

Fig. 4 VSD model developed in DIgSILENT

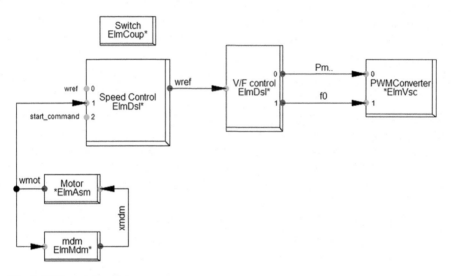

Fig. 5 VSD control model

4.1.1 RMS/EMT Modeling and Simulation with the Motor Breaker in an Open Position

The simulation method to be used should be RMS values (electromechanical transients) for a balanced positive sequence network configuration. The integration step size for electromechanical transients is changed from 0.01 to 0.0001 s to reduce the error of the "system matrix inversion failed." Two events must be set-up, the parameter and switch events. The parameter event is where the start command is defined as shown in Fig. 6. The switching event depends on the speed control parameters.

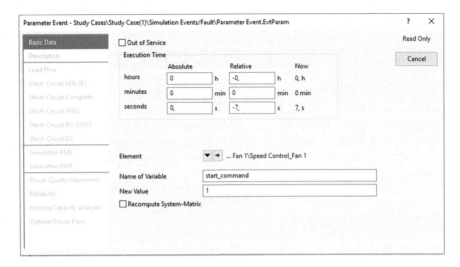

Fig. 6 Parameter event block after simulation of RMS/EMT

The start-up procedure of the induction motor when the motor breaker is open is given in Fig. 7. During the starting process, the transient current is shown in the positive sequence current magnitude waveform.

It is shown in Fig. 8 as the VSD model, the motor breaker _Fan 1 is highlighted with a red circle. The motor breaker _Fan 1 was open and the start-up process must close it before the VSD starts to control the speed. Table 5 shows the output results during the simulation.

4.1.2 RMS/EMT Modeling and Simulation with the Motor Breaker in a Closed Position

Additionally, there is an approach whereby the motor breaker is closed and the fan is off, then the operator decides to start the fan again the motor start-up procedure changes, as well as the resulting output during the simulation. The start-up procedure of the induction motor when the motor breaker is closed is given in Fig. 9. This process only concentrates on the speed control of the induction motor. The VSD model in Fig. 10 shows the flow of current to the induction motor as the motor breaker is in close position. The output results are given in Table 6.

4.2 Fully Rated Converter Wind Turbine Generator (WTG)

Three onshore WTG of 2 MW each was modeled in DIgSILENT, which is 80 m (260 feet) tall according to the datasheet XE93-2 MW [55]. Each wind turbine has three

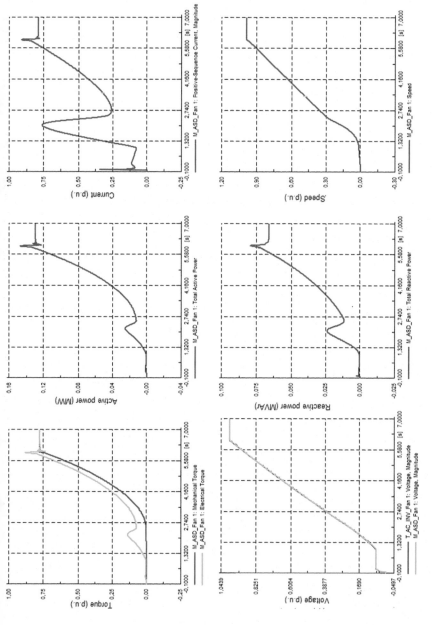

Fig. 7 Motor start-up procedures with Motor Breaker_Fan 1 Open

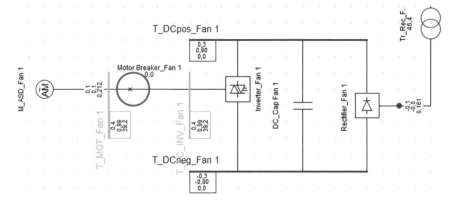

Fig. 8 VSD model when Motor Breaker_Fan 1 opens

Table 5 Result output of the motor starting procedures when Motor Breaker_Fan 1 open

```
●  (t=000:000 ms) ------------------------------------------------------------------
●  (t=000:000 ms) 'Grid\VSD_Model_Fan 1\Speed Control_Fan 1.ElmDsl':
●  (t=000:000 ms) Variable of calculation object: s:start_command set to 1,        0/1    (Motor Start Command OFF/ON)
●  (t=000:055 ms) ------------------------------------------------------------------
●  (t=000:055 ms) 'Grid\VSD_Model_Fan 1\Speed Control_Fan 1.ElmDsl':
●  (t=000:055 ms) Model triggered (t 000:027 ms, event 'ConnectMotor', trigger count 1)!
●  (t=000:027 ms) ------------------------------------------------------------------
●  (t=000:027 ms) 'Grid\Motor Breaker_Fan 1.ElmCoup':
●  (t=000:027 ms) Circuit-Breaker Action: 'Close' - 'All phases'.
```

gigantic blades and a single generator, the total output of this wind farm is 5 MW. The wind turbine is connected at the 0.4 kV LV busbar and supplied by a 2.5 MVA 11/0.4 kV transformer with a Dyn5 vector group. This transformer had a positive sequence and zero sequence impedance of 6% and 3%, respectively. It has copper losses of 2.3 kW. The wind turbine model consists of a harmonic spectrum for better results of the harmonic distortion analysis. The model has the fully rated converter control for the following: WTG 5 MW, slow frequency measurement, PQ controller, PQ LV, voltage measurement, current controller, current measurement, phase-locked loop (PLL), and over frequency power reduction. The schematic diagram of the WTG is shown in Fig. 11, and the wind farm schematic is shown in Fig. 12. A built-in model was used for this research, which is a standard electrical component model that already exists in the DIgSILENT library.

The composite frame model of the *Fully Rated Converter Control* WTG is given in Fig. 13. To ensure that the measuring devices are connected correctly, the measurement devices are connected either to the terminal or to the cubicle that connects the generator to the terminal. The frame description of the *Fully Rated Converter Control* WTG is to reduce the power; in case of electrical over frequency, the active power reduction slot is used. Phase-locked Loop (PLL), PQ, V_{ac}, and I_{ac} are for fast voltage angle, active and reactive power and AC Voltage, and current measuring devices, respectively. For over frequency power reduction, the Slow FrequMeas is used to measure the frequency. The current controller and PQ control are used to calculate

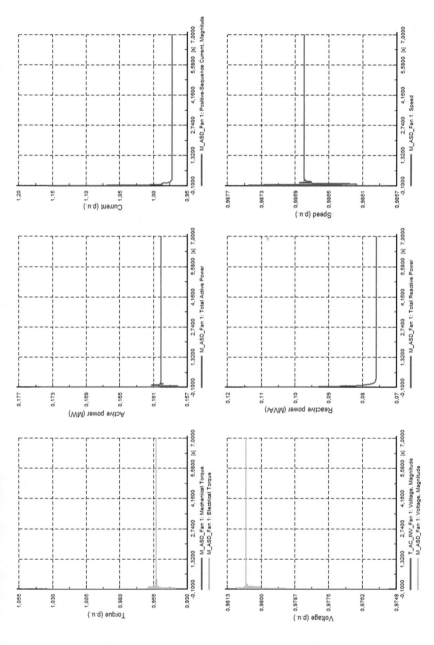

Fig. 9 Motor start-up procedures with Motor Breaker_Fan 1 closed

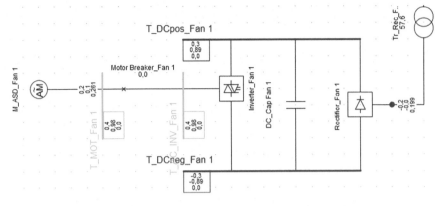

Fig. 10 VSD model when Motor Breaker_Fan 1 is closed

Table 6 Result output of the motor starting procedures when Motor Breaker_Fan 1 closed

```
⊗  (t=000:000 ms) ------------------------------------------------------------------
⊗  (t=000:000 ms) 'Grid\VSD_Model_Fan 1\Speed Control_Fan 1.ElmDsl':
⊗  (t=000:000 ms) Variable of calculation object: s:start_command set to  1,        0/1  (Motor Start Command OFF/ON)
⊗  (t=000:055 ms) ------------------------------------------------------------------
⊗  (t=000:055 ms) 'Grid\VSD_Model_Fan 1\Speed Control_Fan 1.ElmDsl':
⊗  (t=000:055 ms) Model triggered (t 000:027 ms, event 'ConnectMotor', trigger count 1)!
⊗  (t=000:027 ms) Control Switch Event 'ConnectMotor' not possible. Breaker already closed.
```

Fig. 11 Schematic diagram
of the WTG

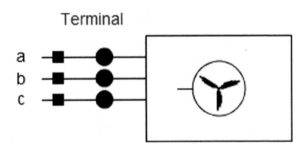

from current reference a voltage signal for the static generator and to control the active and reactive power via the rotor current.

Certain parameters can be changed to adapt the model such as the number of parallel machines, the nominal power required, and nominal AC voltage. It is possible to customize the model. To change the active power reduction during over frequency six parameters need to be changed: frequency which triggers the active power reduction, the frequency which ends the active power reduction, the gradient of the active power reduction, PT1 filter for frequency measurement, gradient limitation for active power reduction, and gradient limitation for active power increase. The over frequency power reduction is a control that uses the frame-block shown in Fig. 14.

The active and reactive power control and current controller are done based on the frame shown in Figs. 15 and 16. These blocks are part of the model in the DIgSILENT.

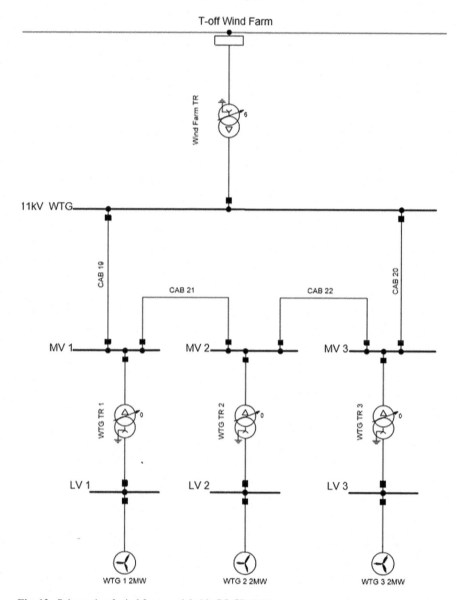

Fig. 12 Schematic of wind farm modeled in DIgSILENT

5 Results and Discussions

This section presents the results and analysis of two different case studies, namely, the network without a wind farm and a network with the integration of Wind Farm. These two case studies are simulated in the DIgSILENT software package. Mine A

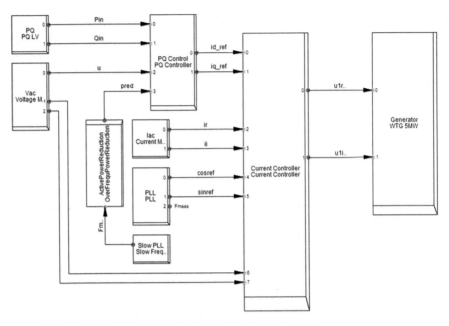

Fig. 13 Composite frame of the Fully Rated Converter Control WTG

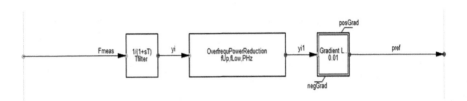

Fig. 14 Over frequency power reduction control

has three-phase 12-pulse rectifiers that are sources of harmonic with the 11th and 13th as dominant harmonic orders. Mine B has many three-phase 6-pulse VSDs that drive the induction motor. These VSDs are sources of harmonics with 5th and 7th as dominant harmonic orders. The solar PV farm and wind farm are RES known to be a harmonic source. This case study is analyzed based on the direction of active power flow, THD, and individual harmonic distortion (HD) and total active power to determine the source of harmonic at the PCC and quantify the contribution. At normal conditions, the active power flows from the grid (utility) and as well as from the RES, in this case, a solar PV farm and wind farm. Due to the complexity of the network, it is a challenge to graphically stand for this network showing the direction of power flow. This network is only analyzed until the 25th harmonic order. The main harmonic source is according to the direction of active power flow, depending on the power at that harmonic frequency. The party with the highest power determines the main harmonic source.

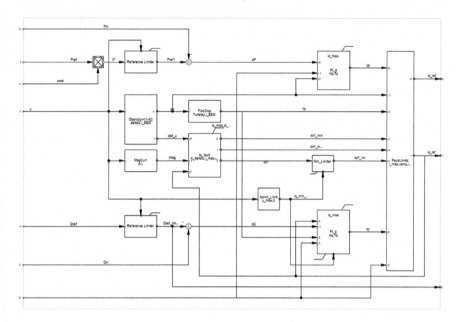

Fig. 15 PQ control frame

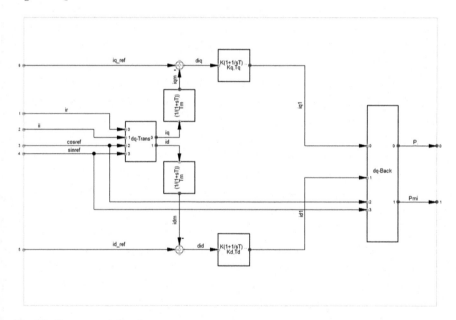

Fig. 16 Current controller frame

5.1   Case Study A

At each harmonic frequency, the harmonic source is found as described in Table 7. Its source in power value is recorded based on the highest value of power in kilowatts to be the harmonic source.

Observations are made according to Table 7 that the direction of active power flow for the 5th, 7th, and 19th harmonic frequency of the solar PV farm flows from the solar plant to the rest of the network. The measured power showed zero, which means that the percentage of the harmonic current is minimal. The results obtained from the case study agree concerning the dominant harmonic each mine has as a theoretical measure. Mine B uses 6-pulse VSDs while Mine A has 12-pulse rectifiers. The active power flow direction shows that customers handle the harmonic seen in this network. The identification of harmonic source per harmonic order is defined as in Table 7.

Table 8 shows that the total power at the PCC is negative which means the harmonic source is from customers. Some power at harmonic frequency is measured at zero, which means that the harmonic power is too small to reach the PCC. This gives a sign of which customer is the main harmonic source measured at the PCC.

Table 9 gives the active power of each customer to quantify the harmonic contribution per harmonic order as well as the total active power to find the contributor of harmonic at the PCC. There are six customers as part of the network, but at the PCC only Mine A and Customer B. Customer B include the solar PV farm, Load 1, Load 2, Load 3, and Mine B. The first approach is to determine the harmonic source between Mine A and Customer B and its quantification, as in Table 9. According to the results obtained in Table 9, the harmonic contributor is Mine A with the highest

Table 7 Determination of harmonic source based on active power direction flow_Case study A

Harmonic frequency	Source of harmonic with power (kW)	Main harmonic source
1st	Utility: 303152.090 kW Solar PV farm: −5785.609 kW	Utility
5th	Mine B: −0.00011 kW Solar PV farm: 0 kW	Mine B
7th	Mine B: −0.00066 kW Solar PV farm: 0 kW	Mine B
11th	Mine A: −2.93826 kW Mine B: −0.56741 kW	Mine A
13th	Mine A: −1.91228 kW Mine B: −0.34593 kW	Mine A
17th	Mine B: −0.00256 kW	Mine B
19th	Mine B: −0.00004 kW Solar PV farm: 0 kW	Mine B
23rd	Mine A: −0.644875 kW	Mine A
25th	Mine A: −0.510115 kW	Mine A

Table 8 Determination of harmonic source by total power at the PCC_ Case study A

Power of individual harmonic frequency	Power (kW)	Sign	Harmonic source
	$P_{Pcc(1)} = -303152.290$	Negative—upstream	Utility
	$P_{Pcc(5)} = -0.00001$	Negative—downstream	Customer—Mine B
	$P_{Pcc(7)} = -0.00007$	Negative—downstream	Customer—Mine B
	$P_{Pcc(11)} = -0.18609$	Negative—downstream	Customer—Mine B
	$P_{Pcc(13)} = -0.02068$	Negative—downstream	Customer—Mine B
	$P_{Pcc(17)} = -0.00093$	Negative—downstream	Customer—Mine B
	$P_{Pcc(19)} = 0$	–	Customer—Mine B
	$P_{Pcc(23)} = -0.18831$	Negative—downstream	Customer—Mine A
	$P_{Pcc(25)} = -0.18167$	Negative—downstream	Customer—Mine A
Total active power	$P_{T\_Pcc} = -303152.868$	Negative	Customer

Table 9 Active power at the harmonic frequency for the customers at the 66 kV PCC_Case study A

Active power (kW)	Mine A	Customer B
P_1	293638.1107	9513.97903
P_5	0.00002	−0.00001
P_7	0.00006	−0.00007
P_{11}	0.28089	−0.18609
P_{13}	0.24534	−0.02068
P_{17}	0.00093	−0.00093
P_{19}	0	0
P_{23}	−0.08192	0.06819
P_{25}	−0.0918	0.09179
P_T calculated	**293638.4642**	9513.93123
TP from DIgSILENT (MW)	**293.637**	9.515

total power at the PCC. Mine A is the one absorbing more power than Customer B at the fundamental frequency.

The second is to quantify the customer contribution within the categories of Customer B as shown in Table 10. It is seen that the solar PV farm generates power at the fundamental frequency while the other loads absorb power. The main contributor

Table 10 Total active power of customers grouped in Customer B_Case study A

Active power (kW)	Load 1	Load 2	Solar PV farm	Load 3	Mine B
P_1	4170.000	960.000	−5785.609	1000.000	9115.704
P_5	0	0	0	0	−0.00011
P_7	0	0	0	0	−0.00066
P_{11}	0.10518	0.01896	0.00283	0.01677	−0.56741
P_{13}	0.03225	0.00578	0.00094	0.00808	−0.34593
P_{17}	0.00004	0.00001	0	0	−0.00256
P_{19}	0	0	0	0	0
P_{23}	0.00121	0.00021	0.00140	0.00192	0.06144
P_{25}	0	0	0.00047	0.00143	0.08846
P_T calculated	4170.139	960.025	-5785.603	1000.028	**9108.863**
TP from DIgSILENT (MW)	4.170	0.960	-5.786	1.000	**9.115**

of harmonic within Customer B is Mine B with the highest total power as well as generating harmonic power as in Table 10.

The harmonic aggregation according to the IEC 61000-3-6 is stipulated in Table 11.

Where α is an exponent being diversity for harmonic summation for the individual harmonic order given by IEC 61000-3-6. Thus the following formulae are used to calculate the total individual harmonics of voltage and current when multiple harmonic sources are connected at one point, as in Eqs. 20 and 21 found in [56]:

$$V_{HDtotal} = \sqrt[\alpha]{\sum V_{HD}^{\alpha}} \tag{20}$$

$$I_{HDtotal} = \sqrt[\alpha]{\sum I_{HD}^{\alpha}} \tag{21}$$

To calculate the THD of voltage and current, the formula used in Eqs. 22 and 23 are given as follows [56]:

$$V_{THDtotal} = \sqrt[\beta]{\sum V_{THD}^{\beta}} \tag{22}$$

Table 11 IEC 61000-3-6 harmonic aggregation method [16]

α	Harmonic order
1	$h < 5$
1.4	$5 < h < 10$
2	$h > 10$

$$I_{THDtotal} = \sqrt[\beta]{\sum I_{THD}^\beta} \qquad (23)$$

where β is the exponent being diversity for harmonic summation for THD of voltage and current. The value of β used in the calculation assumes in [56]. The harmonic summation law is used to determine the total HD and total THD of current for Mine A as the mine has four rectifiers as a harmonic source connected at different busbars. It is also used to calculate the total HD and total THD of voltage as Mine A has double busbars for operational redundancy purposes due to the critical nature of its operation. It is not possible to calculate the summation of HD and THD by straight addition due to phase, time, and spectral diversity as three factors. The combined THD can only possibly be the sum of the individual THDs when each contribution has the same relative spectrum. Table 12 gives the harmonic contribution of each party within a network using the percentage of harmonic distortion as well as defining and quantifying the contribution of each harmonic order. This determines the main contributor of harmonic source based on the THD percentages.

It was found that according to the current total harmonic distortion (ITHD) the harmonic source of this network was Mine A. The individual current harmonic distortion varies depending on the customer's dominant harmonic order. The harmonic source of the individual current harmonic distortion is highlighted in red color in Table 12. The customer(s) who exceeded the harmonic limit per harmonic order is highlighted in blue color, although, in some instances, the customer(s) can be found to be a harmonic contributor and exceeded the limits. The industrial network used range at the I_{sc}/I_L is < 20 as in the current distortion limit for systems rated 120 V through 69 kV [17] in which highlighted that the TDD should not exceed 5%. All customers' ITHDs are within the harmonic limits. It also gives the individual harmonic order limit for the current. The Solar PV individual current HD% of the 23rd and 25th harmonic exceeded the limit. Mine AHD% of the 11th, 13th, 23rd, and 25th exceeded the limit as well.

Table 12 HD and THD of current for the utility and customers_ Case study A

Utility / customers	Individual harmonic distortion (%)								THD
	5th	7th	11th	13th	17th	19th	23rd	25th	
utility	0.008	0.014	1.020	0.933	0.005	0.002	0.338	0.347	1.496
Mine A	0.022	0.034	2.944	2.619	0.018	0.006	1.055	1.019	4.26
Load 1	0.056	0.086	0.594	0.520	0.036	0.045	0.059	0.002	0.859
Load 2	0.051	0.077	0.526	0.459	0.031	0.040	0.052	0.002	0.758
Solar PV	0.367	0.522	0.987	1.258	0.173	0.313	1.325	0.778	2.872
Load 3	0.010	0.019	0.417	0.311	0.010	0.004	0.139	0.119	0.588
Mine B	0.306	0.600	1.197	0.950	0.287	0.269	0.543	0.477	1.931

▐ Harmonic Source Contributor

▐ Exceed the harmonic limits

▐ Harmonic Source Contributor & Exceed the harmonic limit

Table 13 HD and THD of voltage at different busbars_ Case study A

Busbar name	Individual Harmonic Distortion (%)								THD
	5th	7th	11th	13th	17th	19th	23rd	25th	(%)
PV Plant 33kV	0.10	0.20	2.02	2.07	0.18	0.26	0.43	0	4.53
33kV Busbar	0.10	0.20	2.02	2.06	0.18	0.26	0.40	0.01	4.51
PCC 66kV – Customer B 66kV	0.02	0.05	1.73	1.50	0.06	0.02	1.11	1.02	3.55
Mine A – No. 1 66kV Busbar	0.01	0.01	2.18	1.95	0.01	0	1.52	1.40	4.10
Mine A – No. 2 66kV Busbar	0.01	0.01	2.18	1.95	0.01	0	1.51	1.39	4.07
Mine A – No. 1 and 2 Summation	0.02	0.02	3.08	2.76	0.01	0	2.14	1.97	5.78
Mine B 66kV	0.02	0.05	1.72	1.49	0.06	0.03	1.11	1.02	3.62
66kV Node	0.02	0.05	1.72	1.49	0.06	0.03	1.11	1.03	3.65
PCC 66kV Busbar	0.01	0.01	1.74	1.48	0.01	0	1.21	1.06	3.25

▮ Harmonic Source Contributor

▮ Exceed the harmonic limits

▮ Harmonic Source Contributor & Exceed the harmonic limit

Table 13 gives the individual and THD of the voltage at different points within the network. It is seen that according to the voltage total harmonic distortion(VTHD) the Mine A—No. 1 and 2 Summation is a harmonic source in the network. The 11th harmonic order of Mine A—No. 1 and 2 Summation has exceeded the harmonic limit prescribed by the international standard.

The individual and THD of the voltage are within the prescribed limit set by IEEE Std. 519-2014. This industrial network range is between $1\,kV < V \leq 69\,kV$, which said that the VTHD should not exceed 5% while the HD of the voltage should not exceed 3%. Thus, the recorded VTHD % and 11th individual voltage HD % exceeded the summation of the Mine A—No. 1 and 2 only. For the PCC 66 kV busbar, the VTHD % is still within the prescribed limit, thus for this network, the harmonic filters are working accordingly.

Besides, the main harmonic contributor of this case study is Mine A this is seen by the total power at the PCC as well as the percentages of the VTHD and ITHD. The direction of active power is a good indication to show the flow of power within a network at different frequencies. The individual harmonic order contributor has been found and this is a good corrective measure to decide on the harmonic mitigation depends on the dominant harmonic order of the network.

5.2   Case Study B

With the introduction of the wind power farm, the harmonic source detection is analyzed on the industrial network and saw the differences. The harmonic source for the individual harmonic frequency is found and tabulated in Table 14.

The direction of active power flow at different frequencies is seen and the fundamental frequency or the 1st harmonic frequency the power is from the utility and the RES (solar PV and wind farm). The 5th harmonic contributor is a wind farm, thus

Table 14 Determination of harmonic source based on active power direction flow_ Case study B

Harmonic frequency	Source of harmonic with power (kW)	Main harmonic source
1st	Utility: 297219.429 kW Solar PV farm: −5793.753 kW Wind farm: −5976.465 kW	Utility
5th	Mine B: −0.00010 kW Solar PV farm: 0 kW Wind farm: −0.00014 kW	Wind farm
7th	Mine B: −0.00060 kW Solar PV farm: 0 kW Wind farm: −0.00003 kW	Mine B
11th	Mine A: −2.94711 kW Mine B: −0.56437 kW	Mine A
13th	Mine A: −1.92193 kW Mine B: −0.34151 kW	Mine A
17th	Mine B: −0.00134 kW Wind farm: −0.00010 kW	Mine B
19th	Mine B: −0.00004 kW Solar PV farm: 0 kW Wind farm: 0 kW	Mine B
23rd	Mine A: −0.553 kW	Mine A
25th	Mine A: −0.48331 kW	Mine A

this was only the harmonic frequency that changes the harmonic contributor, the rest of the harmonic frequency is like case study A only the value of active power that changes significantly. Table 15 shows that the total power at the PCC is negative which means the harmonic source is from customers. This customer must be found based on the percentage of the ITHDs.

Table 15 gives the active power of each customer to quantify the harmonic contribution per harmonic order as well as the total active power to find the contributor of harmonic at the PCC. There are seven customers as part of the network, but at the PCC only Mine A and Customer B. Customer B includes the wind farm, solar PV farm, Load 1, Load 2, Load 3, and Mine B. The first approach is to determine the harmonic source between Mine A and Customer B and its quantification, as shown in Table 16. According to the results obtained in Table 18, the harmonic contributor is Mine A with the highest total power at the PCC. Mine A is the one absorbing more power than Customer B at the fundamental frequency.

The second is to quantify the customer contribution within the categories of Customer B as shown in Table 17. It is seen that wind farm and solar PV farm generates power at the fundamental frequency while the other loads absorb power. The main contributor of harmonic within Customer B is Mine B with the highest total power as well as generating harmonic power as in Table 17.

Table 18 gives the harmonic contribution of each party within a network using the percentage of harmonic distortion as well as defining and quantifying the contribution

Table 15 Determination of harmonic source by total power at the PCC_ Case study B

Power of individual harmonic frequency	Power (kW)	Sign	Harmonic source
	$P_{Pcc(1)} = -297219.429$	Negative—upstream	Utility
	$P_{Pcc(5)} = -0.00005$	Negative—downstream	Customer—wind farm
	$P_{Pcc(7)} = -0.00007$	Negative—downstream	Customer—Mine B
	$P_{Pcc(11)} = -0.12738$	Negative—downstream	Customer—Mine B
	$P_{Pcc(13)} = -0.06897$	Negative—downstream	Customer—Mine B
	$P_{Pcc(17)} = -0.00026$	Negative—downstream	Customer—Mine B
	$P_{Pcc(19)} = 0$	–	Customer—Mine B
	$P_{Pcc(23)} = -0.20167$	Negative—downstream	Customer—Mine A
	$P_{Pcc(25)} = -0.17123$	Negative—downstream	Customer—Mine A
Total active power	$P_{T\_Pcc} = -297219.999$	Negative	Customer

Table 16 Active power at the harmonic frequency for the customers at the 66 kV PCC_Case study B

Active power (kW)	Mine A	Customer B
P_1	293699.925	3519.504
P_5	0.00006	−0.00005
P_7	0.00006	−0.00007
P_{11}	0.27325	−0.12738
P_{13}	0.23248	−0.06897
P_{17}	0.00026	−0.00026
P_{19}	0	0
P_{23}	−0.17101	0.14155
P_{25}	−0.08811	0.07974
P_T calculated	**293699.940**	3519.529
TP from DIgSILENT (MW)	**293.699**	3.520

of each harmonic order. This determines the main contributor of harmonic source based on the ITHD percentages.

Based on the percentage of ITHD the wind farm is the contributor of harmonic distortion within the network, this can be because the mining areas and solar PV farm have harmonic filters installed as part of their mitigation. The wind farm has no mitigation factors installed as part of their plants. The individual current harmonic

Table 17 Total active power of customers grouped in Customer B_Case study B

Active power (kW)	Load 1	Load 2	Wind farm	Solar PV farm	Load 3	Mine B
P_1	4170.000	959.999	−5976.465	−5785.980	1000.000	9115.600
P_5	0	0	−0.00014	0	0	−0.00010
P_7	0	0	−0.00003	0	0	−0.00060
P_{11}	0.10833	0.01953	0.00037	0.00293	0.01715	−0.56437
P_{13}	0.03466	0.00621	0.00049	0.00103	0.00842	−0.34151
P_{17}	0.00002	0	−0.00010	0	0	−0.00134
P_{19}	0	0	0	0	0	−0.00004
P_{23}	0.00012	0.00002	0.03176	0.00014	0.00021	0.00725
P_{25}	0	0	0.00513	0.00027	0.00081	0.06394
P_T calculated	4170.143	960.025	−5976.428	−5785.976	1000.027	**9114.763**
TP from DIgSILENT (MW)	4.170	0.960	−5.976	−5.786	1.000	**9.115**

Table 18 HD and THD of current for the utility and customers_ Case study B

Utility / customers	Individual harmonic distortion (%)								THD
	5th	7th	11th	13th	17th	19th	23rd	25th	
utility	0.014	0.016	1.047	0.961	0.007	0.003	0.285	0.332	1.517
Mine A	0.023	0.024	1.486	1.337	0.015	0.004	0.683	0.567	2.218
Load 1	0.058	0.087	0.605	0.534	0.041	0.049	0.026	0.002	0.865
Load 2	0.053	0.078	0.535	0.471	0.036	0.043	0.023	0.001	0.764
Wind farm	0.794	0.397	0.875	1.026	0.338	0.253	9.093	2.979	9.836
Solar PV	0.367	0.523	1.007	1.296	1.200	0.339	0.594	0.609	2.528
Load 3	0.014	0.020	0.424	0.320	0.012	0.004	0.064	0.093	0.574
Mine B	0.315	0.604	1.225	0.981	0.288	0.270	0.366	0.417	1.904

▉ Harmonic Source Contributor

▉ Exceed the harmonic limits

▉ Harmonic Source Contributor & Exceed the harmonic limit

distortion that exceeded the harmonic limit is 23rd for Mine A, the 23rd and 25th harmonic of the wind farm, and the 25th harmonic order of the solar PV farm. The wind farm percentages of ITHD exceeded the limit of 5%.

Table 19 gives the individual and THD of the voltage at different points within the network. It was seen that according to the VTHD for the Mine A—No. 1 and 2 Summation of 11th has exceeded the harmonic limit. The 11 kV WTG busbar exceeded the harmonic limit of the 23rd and 25th harmonic orders. The percentage of the VTHD of the 11 kV WTG busbar is the highest within the network and

Table 19 HD and THD of voltage at different busbars_Case study B

Busbar name	Individual Harmonic Distortion (%)								THD
	5th	7th	11th	13th	17th	19th	23rd	25th	(%)
PV Plant 33kV	0.10	0.20	2.06	2.13	0.21	0.28	0.19	0	4.33
33kV Busbar	0.10	0.20	2.05	2.12	0.21	0.28	0.18	0.01	4.31
11kV WTG	0.24	0.18	2.32	2.30	0.36	0.29	11.48	3.45	12.50
PCC 66kV – Customer B 66kV	0.02	0.03	1.77	1.53	0.04	0.02	0.53	0.82	3.12
Mine A – No. 1 66kV Busbar	0.01	0.01	2.19	1.96	0.01	0	1.41	1.37	4.05
Mine A – No. 2 66kV Busbar	0.01	0.01	2.19	1.96	0.01	0	1.40	1.36	4.03
Mine A – No. 1 and 2 Summation	0.02	0.02	3.10	2.77	0.02	0	1.99	1.93	5.71
Mine B 66kV	0.03	0.05	1.74	1.53	0.07	0.03	0.49	0.80	3.32
66kV Node	0.03	0.05	1.74	1.53	0.07	0.03	0.49	0.80	3.32
PCC 66kV Busbar	0.01	0.01	1.76	1.50	0.02	0.01	1.00	0.99	3.15

▮ Harmonic Source Contributor

▮ Exceed the harmonic limits

▯ Harmonic Source Contributor & Exceed the harmonic limit

exceeded the harmonic limit. The percentage of the VTHD of Mine A—No. 1 and 2 Summation exceeded the harmonic limit.

Thus, harmonic mitigation is needed for the wind farm as is the only plant without the harmonic filters installed. This harmonic filter needs to mitigate the dominant harmonic order which is the 23rd and 25th. The harmonic filter for the wind farm is designed according to the stages specified in [13]. The damped filter single tuned is selected for this design because is used to eliminate the high order harmonics which is the 23rd harmonics. This filter also called a high pass harmonic filter.

The harmonic filters are designed according to the following formulae: the reactive power for the capacitor is determined based on the power factor correction.

Step 1:

$$Q_c = P(\tan \delta_1 - \tan \delta_2) \tag{24}$$

Step 2:

$$X_c = \frac{kV^2}{Q_c} \tag{25}$$

Step 3:

$$X_L = \frac{X_c}{hr^2} \tag{26}$$

Step 4:

$$X_n = \sqrt{X_c \times X_L} \tag{27}$$

Step 5:

$$Q = 0.5 < Q < 5, Q = 5 \tag{28}$$

Step 6:

$$R = \frac{X_n}{Q} \tag{29}$$

Step 7:

$$L = \frac{X_L}{2\pi f_1} \tag{30}$$

Step 8:

$$C = \frac{1}{2\pi f_1 X_c} \tag{31}$$

Step 9:

$$Q_{Filter} = \frac{kV^2}{X_c - X_L} \tag{32}$$

The harmonic at the 11 kV WTG busbar is eliminated and this scenario caused the case study B to be repeated with the installed harmonic filter and analyze the industrial network.

5.3 Case Study C

Wind power farm has installed harmonic filter as part of the plant to mitigate the harmonic content produced by the plant. In Table 20 the harmonic source for the individual harmonic frequency is seen.

The direction of active power flow at different frequencies is seen and the fundamental frequency or the 1st harmonic frequency the power is from the utility and the RES (solar PV and wind farm). The results of the case study C is like the results obtained for case study B only some value of the power at the different harmonic frequency that changes.

The downstream of the network is a contributor as shown in Table 21 when it shows the total power at the PCC which is negative. The active power of each customer connected at the downstream is obtained to quantify the harmonic contributor of the individual harmonic order and the main contributor of the network.

From the active power in Table 22, Mine A is the main contributor to this harmonic in this network but this must be seen again on the percentage of the ITHDs and

Table 20 Determination of harmonic source based on active power direction flow_ Case study C

Harmonic frequency	Source of harmonic with power (kW)	Main harmonic source
1st	Utility: 297233.406 kW Solar PV farm: −5785.997 kW Wind farm: −5975.757 kW	Utility
5th	Mine B: −0.00010 kW Solar PV farm: 0 kW Wind farm: −0.00014 kW	Wind farm
7th	Mine B: −0.00060 kW Solar PV farm: 0 kW Wind farm: −0.00003 kW	Mine B
11th	Mine A: −2.98836 kW Mine B: −0.56846 kW	Mine A
13th	Mine A: −2.00791 kW Mine B: −0.34510 kW	Mine A
17th	Mine B: −0.00094 kW Wind farm: −0.00011 kW	Mine B
19th	Mine B: −0.00003 kW Solar PV farm: 0 kW	Mine B
23rd	Mine A: −0.69668 kW	Mine A
25th	Mine A: −0.55736 kW	Mine A

Table 21 Determination of harmonic source by total power at the PCC_ Case study C

Power of individual harmonic frequency	Power (kW)	Sign	Harmonic source
	$P_{Pcc(1)} = -297233.406$	Negative—upstream	utility
	$P_{Pcc(5)} = -0.00006$	Negative—downstream	Customer—Wind farm
	$P_{Pcc(7)} = -0.00007$	Negative—downstream	Customer—Mine B
	$P_{Pcc(11)} = -0.14035$	Negative—downstream	Customer—Mine B
	$P_{Pcc(13)} = -0.14521$	Negative—downstream	Customer—Mine B
	$P_{Pcc(17)} = -0.00023$	Negative—downstream	Customer—Mine B
	$P_{Pcc(19)} = 0$	–	Customer—Mine B
	$P_{Pcc(23)} = -0.34499$	Negative—downstream	Customer—Mine A
	$P_{Pcc(27)} = -0.31386$	Negative—downstream	Customer—Mine A
Total active power	$P_{T\_Pcc} = -297234.351$	Negative	Customer

Table 22 Active power at the harmonic frequency for the customers at the 66 kV PCC_Case study C

Active power (kW)	Mine A	Customer B
P_1	293713.301	3520.105
P_5	0.00006	−0.00006
P_7	0.00006	−0.00007
P_{11}	0.15243	−0.14035
P_{13}	−0.06715	−0.03251
P_{17}	0.00022	−0.00023
P_{19}	0	0
P_{23}	−0.34499	0.31294
P_{25}	−0.31386	0.29264
P_T calculated	**293712.728**	3520.537
TP from DIgSILENT (MW)	**293.713**	3.521

VTHDs. It is important to analyze the customers categorized in Customer B, to quantify which customer is the main contributor within categories B.

Thus, it was found in Table 23 that the main contributor of harmonic within category Customer B is Mine B because most of the harmonic frequency is the one injecting harmonic current within the network.

Table 24 gives the harmonic contribution of each party within a network using the percentage of harmonic distortion as well as defining and quantifying the contribution of each harmonic order. This determines the main contributor of harmonic source based on the ITHD percentages. Based on the percentage of ITHD the wind farm is the contributor of harmonic distortion within the network even after the harmonic

Table 23 Total active power of customers grouped in Customer B_Case study C

Active power (kW)	Load 1	Load 2	Wind farm	Solar PV farm	Load 3	Mine B
P_1	4170.000	959.999	−5975.757	−5785.997	1000.000	9115.597
P_5	0	0	−0.00014	0	0	−0.00010
P_7	0	0	−0.00003	0	0	−0.00060
P_{11}	0.11193	0.02018	0.09834	0.00306	0.01764	−0.56846
P_{13}	0.03831	0.00687	0.23549	0.00116	0.00901	−0.34510
P_{17}	0.00001	0	−0.00011	0	0	−0.00094
P_{19}	0	0	0	0	0	−0.00003
P_{23}	0.000105	0.000019	0.22438	0.00123	0.00168	0.07276
P_{25}	0	0	0.18734	0.00043	0.00130	0.10016
P_T calculated	4170.151	960.026	−5975.012	−5785.991	1000.030	**9114.855**
TP from DIgSILENT (MW)	4.170	0.960	−5.975	−5.786	1.000	**9.115**

Table 24 HD and THD of current for the utility and customers_ Case study C

Utility / customers	Individual harmonic distortion (%)								THD
	5th	7th	11th	13th	17th	19th	23rd	25th	
utility	0.015	0.016	1.053	0.972	0.007	0.002	0.338	0.349	1.543
Mine A	0.024	0.024	1.473	1.315	0.016	0.003	0.531	0.520	2.142
Load 1	0.058	0.087	0.614	0.553	0.040	0.045	0.056	0.002	0.882
Load 2	0.053	0.078	0.543	0.488	0.038	0.039	0.049	0.002	0.780
Wind farm	0.815	0.421	1.638	2.420	0.638	0.106	0.926	0.586	3.747
Solar PV	0.368	0.523	1.022	1.343	0.197	0.309	1.253	0.749	2.808
Load 3	0.014	0.020	0.431	0.331	0.011	0.004	0.131	0.115	0.599
Mine B	0.315	0.604	1.243	1.011	0.294	0.270	0.526	0.465	1.977

◼ Harmonic Source Contributor

◼ Exceed the harmonic limits

▦ Harmonic Source Contributor & Exceed the harmonic limit

mitigation installed to be part of the network. Mine B only the harmonic source contributor of the 7th harmonic order. Solar PV is the harmonic source contributor of the 19th, 23rd, and 25th harmonic order, thus, the 23rd and 25th harmonic order exceeded the harmonic order. Wind farm as a main harmonic order based on the ITHD percentage, harmonic source contributor of the 5th, 13th, and 17th, as well as the harmonic orders recorded with harmonic limits exceeded, is the 13th and 23rd.

Table 25 gives the individual and THD of the voltage at different points within the network. It was seen that according to the VTHD for Mine A—No. 1 and 2 Summation of the 11th has exceeded the harmonic limit. The percentage of the VTHD also exceeded harmonic limits and is the highest value recorded for the network. The 11 kV WTG busbar exceeded the harmonic limit of the 13th harmonic order after the installation of the harmonic filter.

Table 25 HD and THD of voltage at different busbars_Case study C

Busbar name	Individual Harmonic Distortion (%)								THD (%)
	5th	7th	11th	13th	17th	19th	23rd	25th	
PV Plant 33kV	0.10	0.20	2.09	2.20	0.21	0.26	0.40	0	4.35
33kV Busbar	0.10	0.20	2.08	2.19	0.21	0.25	0.38	0.01	4.33
11kV WTG	0.25	0.19	2.83	3.39	0.66	0.12	0.51	0.60	4.83
PCC 66kV – Customer B 66kV	0.02	0.03	1.79	1.59	0.05	0.01	1.10	1.00	3.33
Mine A – No. 1 66kV Busbar	0.01	0.01	2.20	1.97	0.01	0	1.51	1.39	4.10
Mine A – No. 2 66kV Busbar	0.01	0.01	2.19	1.97	0.01	0	1.50	1.38	4.07
Mine A – No. 1 and 2 Summation	0.02	0.02	3.10	2.79	0.02	0	2.13	1.96	5.78
Mine B 66kV	0.03	0.05	1.77	1.59	0.07	0.03	1.05	0.98	3.51
66kV Node	0.03	0.05	1.77	1.59	0.07	0.03	1.05	0.98	3.51
PCC 66kV Busbar	0.01	0.01	1.76	1.52	0.02	0	1.18	1.05	3.24

◼ Harmonic Source Contributor

◼ Exceed the harmonic limits

▦ Harmonic Source Contributor & Exceed the harmonic limit

5.4 Comparison of Different Case Studies

The three case studies results were compared to draw the conclusion based on the results obtained in the different case studies namely:

(a) Case study A—Industrial network with solar PV farm
(b) Case study B—Wind farm introduced on the industrial network
(c) Case study C—Additional harmonic filter modeled part of the wind farm

The main three categories of information are compared, namely,

(a) *Total power at the PCC to determine whether the harmonic source contributor is from the upstream or downstream the PCC:*

For all three case studies, it was noticed that the harmonic source contributor of the network is from the downstream of the PCC in other word is from the customer's side as seen in Table 26. Besides, DIgSILENT is a useful software that gives the direction of active power flow through the arrows this enabled to capture of the power of each customer who is injecting harmonic at that specific harmonic frequency and identify who is the harmonic contributor depends on the power value as described in Table 27.

Table 26 Comparison of the total power at the PCC

Case studies	Harmonic source		
	Total active power (kW)	Sign	Harmonic source
Case study A	$P_{T\_Pcc} = -303152.868$	Negative	Customer
Case study B	$P_{T\_Pcc} = -297219.999$	Negative	Customer
Case study C	$P_{T\_Pcc} = -297234.351$	Negative	Customer

Table 27 Main source of harmonic at each harmonic frequency based on the direction of active power flow arrow

Harmonic frequency	Main harmonic source		
	Case study A	Case study B	Case study C
1st	Utility	Utility	Utility
5th	Mine B	Wind farm	Wind farm
7th	Mine B	Mine B	Mine B
11th	Mine A	Mine A	Mine A
13th	Mine A	Mine A	Mine A
17th	Mine B	Mine B	Mine B
19th	Mine B	Mine B	Mine B
23rd	Mine A	Mine A	Mine A
25th	Mine A	Mine A	Mine A

(b) *The percentage of the ITHDs of the different customer within the network to determine which customer is injecting the highest current distortion into the network:*

Based on the percentage of the ITHDs in Table 28 of different customers form part of the network it was found that Mine A was the main contributor for harmonic for case study A. For case studies B and C, the wind farm is the main harmonic source contributor of this network. Only in case study B where the percentage of ITHD of the harmonic source contributor exceeded the limit under the standards.

(c) *The percentage of the VTHDs of different busbars within the network to determine which busbar within the network is affected by the harmonic which are part of the network:*

Based on the comparison shown in Table 29 shows that the busbar that is affected more by the harmonic distortion is the Mine A—No. 1 and 2 66 kV Summation in all case studies. The 11 kV WTG busbar was affected only before the installation of the harmonic filter. The concern about the PCC is that the harmonic distortion that reaches the PCC is within the harmonic limit for all the case studies.

Table 28 Comparison of the percentage for ITHD for the utility and customers

Case studies	ITHD (%)						
	Customers						
	Mine A	Load 1	Load 2	Wind farm	Solar PV farm	Load 3	Mine B
Case study A	4.26	0.859	0.758	N/A	2.872	0.588	1.931
Case study B	2.218	0.865	0.764	9.836	2.528	0.574	1.904
Case study C	2.142	0.882	0.780	3.747	2.808	0.599	1.977

Table 29 Comparison of the percentage for VTHD at different busbars

Busbar name	Case studies		
	Case study A	Case study B	Case study C
PV Plant 33kV	4.53	4.33	4.35
33kV Busbar	4.51	4.31	4.33
11kV WTG	N/A	12.50	4.83
PCC 66kV – Customer B 66kV	3.55	3.12	3.33
Mine A – No. 1 66kV Busbar	4.10	4.05	4.10
Mine A – No. 2 66kV Busbar	4.07	4.03	4.07
Mine A – No. 1 and 2 Summation	5.78	5.71	5.78
Mine B 66kV	3.62	3.32	3.51
66kV Node	3.65	3.32	3.51
PCC 66kV Busbar	3.25	3.15	3.24

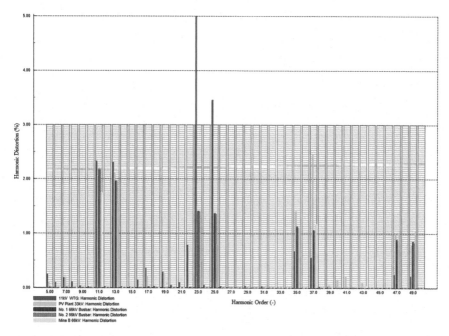

Fig. 17 Individual voltage harmonic spectrum for case study B

5.5 Waveforms Comparison of the Two Case Studies B and C

The graphical representation that forms part of this chapter is only for case study B and Case study C with wind farm part of the industrial network.

5.5.1 Individual Harmonic Distortion

Figure 17 and Fig. 18 give the individual voltage harmonic spectrum for the two case studies B and C, respectively. The individual voltage harmonic distortion should not exceed 3%. The individual voltage harmonic distortion for case study A shows that the 23rd and 25th exceeded the harmonic limits prescribed by IEEE 519-2014.

Based on case study C when the harmonic filter is part of the wind farm plant, the only harmonic order that exceeds the limit is the 13th. The 23rd and 25th harmonic orders are well-tuned by the harmonic filter.

5.5.2 Voltage Waveform at the 11 kV WTG Busbar

The comparison of the voltage waveforms for two case studies is discussed by pointing out the issues causing the waveform shapes. The known waveform of

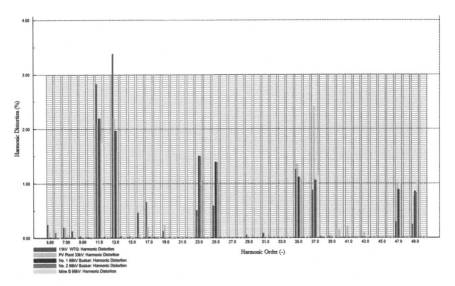

Fig. 18 Individual voltage harmonic spectrum for case study C

voltage should be sinusoidal; however, in today's network, the voltage waveforms are distorted because of harmonic distortion. Figure 19 gives the voltage waveform

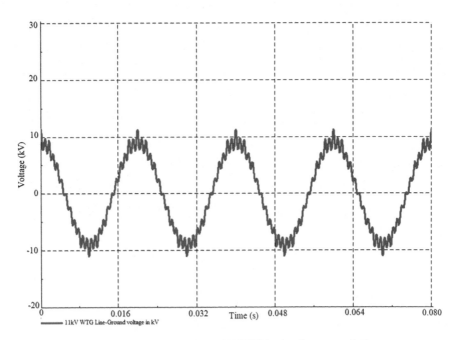

Fig. 19 Voltage distortion waveform at the 11 kV WTG busbar for case study B

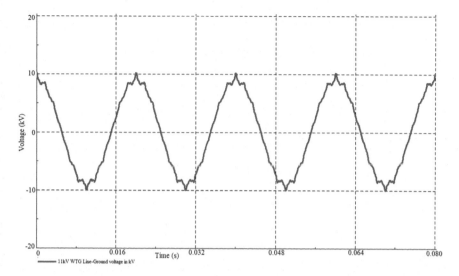

Fig. 20 Voltage distortion waveform at the 11 kV WTG busbar for case study C

for case study A. The waveform is distorted; as the network experiences harmonic distortion which is exceeded the limit. The 11 kV WTG is a selected busbar because is where the WTGs are connected.

The waveform for case study B is too distorted compared to the voltage waveform for case study C shown in Fig. 20. This shows how the harmonic filter reduces the distortion level within the busbars.

6 Conclusions

The analysis provided in this chapter confirms that the wind energy systems can inject harmonic current into the network, and this can contribute to the THD at the point of connection. DIgSILENT software is used to model wind turbines and VSDs. The modified active power direction flow method of harmonic source detection is used to analyze the source of harmonic distortion for the network at three different case studies. The results of the three case studies of a typical Southern Africa industrial network, along with voltage waveforms and individual harmonic distortion, are compared to give a better understanding of the harmonic contribution from different sides of the network. The design of a harmonic mitigation filter for the wind farm is included in this chapter where a damped high pass single tuned harmonic filter is used to eliminate the 23rd harmonic order which is the dominant harmonics within the wind farm.

References

1. Ibrahem H, Yehia DM, Azmy AM (2019) Effect of wind energy penetration scenario on power quality and power losses of distribution networks. In: 2019 IEEE conference on power electronics and renewable energy, pp 1–6

2. Golovanov N, Lazaroiu GC, Roscia M, Zaninelli D (2013) Power quality assessment in small scale renewable energy sources supplying distribution systems. Energies 6(2):634–645

3. Liang S, Hu Q, Lee W (2012) A survey of harmonic emissions of a commercially operated wind farm. IEEE Trans Ind Appl 48(3):1115–1123

4. Vilchez E, Stenzel J (2009) Comparison of wind energy integration into a 110 kV and a 380 kV transmission system—impact on power quality of MV and LV networks. In: Asia-Pacific power and energy engineering conference, pp 1–6

5. Vilchez E, Stenzel J (2008) Wind energy integration into 110 kV system—impact on power quality of MV and LV networks. In: 2008 IEEE/PES transmission and distribution conference and exposition, pp 1–6

6. Preciado V, Madrigal M, Muljadi E, Gevorgian V (2015) Harmonics in a wind power plant. In: IEEE power and energy society general meeting, vol 2015-Sept, no April 2015

7. Tentzerakis S, Paraskevopoulou N, Papathanassiou S, Papadopoulos P (2008) Measurement of wind farm harmonic emissions. In: PESC record—IEEE annual power electronics specialists conference, pp 1769–1775

8. Johnson DO, Hassan KA (2016) Issues of power quality in electrical systems. Int J Energy Power Eng 5(4):148–154

9. Dugan RC, McGranaghan MF, Santoso S, Beaty HW (2003) Electrical power systems quality, 2nd ed, vol 1

10. Islam KMS, Samra AH (1997) Identification of harmonic sources in power distribution systems. In: Proceedings IEEE SOUTHEASTCON 97 engineering the new century, pp 301–303

11. Mohod SW, Aware MV (2013) Power quality and grid code issues in wind energy conversion system. In: An update on power quality, pp 21–36

12. Ruiz-Cortés M, Milanés-Montero MI, Barrero-González F, Romero-Cadaval E (2015) Analysis of causes and effects of harmonic distortion in electric power systems and solutions to comply with international standards regarding power quality. In: Doctoral conference on computing, electrical and industrial systems, pp 357–364

13. Wakileh GJ (2001) Power systems harmonics: fundamentals, analysis and filter design. Springer, New York

14. Dao T, Phung BT (2018) Effects of voltage harmonic on losses and temperature rise in distribution transformers. IET Gener Transm Distrib 12(2):347–354

15. Dao T, Halim HA, Liu Z, Phung BT (2017) Voltage harmonic effect on losses in distribution transformers. In: 2016 International conference on smart green technology in electrical and information systems: advancing smart and green technology to build smart society, no October, pp 27–32

16. IEC Std. 61000-3-6 (2013) Electromagnetic compatibility (EMC)—Part 3-6: Limits—assessment of emission limits for the connection of distorting installations to MV, HV and EHV power systems

17. IEEE Std 519-2014 (2014) IEEE recommended practice and requirements for harmonic control in electric power systems. IEEE Std 519-2014 (Revision IEEE Std 519-1992), pp 1–29

18. NRS 048-4:2009 (2009) Electricity supply—quality of supply: Part 4 : Application practices for licensees. NRS 048-42009, pp 1–57

19. Yuvaraj V, Raj EP, Mowlidharan A, Thirugnanamoorthy L (2011) Power quality improvement for grid connected wind energy system using FACTS device. In: Proceedings of the joint 3rd international workshop on nonlinear dynamics and synchronization, pp 301–307

20. Negumbo R (2009) Analysis for electrical energy and overall efficiency in distribution network with harmonic distortion. Cape Peninsula University of Technology, Cape Town

21. Bezuidenhout SL (2003) A comparative study into the application of the NRS 048, IEEE 519-1992 and IEC 61000-3-2 on harmonic apportioning in a discrimitive tariff. North-West University
22. Atkinson-Hope G (2004) Relationship between harmonics and symmetrical components. Int J Electr Eng Educ 41(2):93–104
23. Sinvula R, Abo-Al-Ez KM, Kahn MT (2019) Harmonic source detection methods: a systematic literature review. IEEE Access 7:74283–74299
24. Swart PH, Case MJ, Van Wyk JD (1994) On techniques for localization of sources producing distortion in electric power networks. Eur Trans Electr Power 4(6):485–489
25. Lin W, Lin C, Tu K, Wu C (2005) Multiple harmonic source detection and equipment identification with cascade correlation network. IEEE Trans Power Deliv 20(3):2166–2173
26. Stevanović D, Petković P (2011) A new method for detecting source of harmonic pollution at grid. In: Proceedings of 16th international symposium on power electronics, vol 26, no 28.10, pp 1–4
27. Xu F, Yang H, Zhao J, Wang Z, Liu Y (2018) Study on constraints for harmonic source determination using active power direction. IEEE Trans Power Deliv 33(6):2683–2692
28. Malekian K (2015) A novel approach to analyze the harmonic behavior of customers at the point of common coupling. In: 2015 9th international conference on compatibility and power electronics, pp 31–36
29. Cai G, Wang L, Yang D, Sun Z, Wang B (2019) Harmonic detection for power grids using adaptive variational mode decomposition. Energies 12(2):232
30. Li C-S, Bai Z-X, Xiao X-Y, Liu Y-M, Zhang Y (2016) Research of harmonic distortion power for harmonic source detection. In: 2016 17th international conference on harmonics and quality of power, pp 126–129
31. Gül Ö, Gündoğdu T (2015) Harmonic contributions of utility and customer based on load model using field measurements. J Power Energy Eng 3(5):11–28
32. Lin R, Xu L, Zheng X (2018) A method for harmonic sources detection based on harmonic distortion power rate. IOP Conf Seri: Mater Sci Eng 322(7):1–7
33. Peterson B, Rens J, Botha MG, Desmet J (2017) On harmonic emission assessment: a discriminative approach. SAIEE Africa Res J 108(4):165–173
34. Karimzadeh F, Esmaeili S, Hosseinian SH (2016) Method for determining utility and consumer harmonic contributions based on complex independent component analysis. IET Gener Transm Distrib 10(2):526–534
35. Karimzadeh F, Esmaeili S, Hosseinian SH (2015) A novel method for noninvasive estimation of utility harmonic impedance based on complex independent component analysis. IEEE Trans Power Deliv 30(4):1843–1852
36. Zhao X, Yang H (2015) Method of calculating system-side harmonic impedance based on FastICA. Autom Electr Power Syst 39(23):139–144
37. Zhao X, Yang H (2016) A new method to calculate the utility harmonic impedance based on FastICA. IEEE Trans Power Deliv 31(1):381–388
38. Jiang W, Su N, Ding L, Qiu S (2016) Assessment method of harmonic emission level based on the improved weighted support vector machine regression. In: Proceedings of the 2015 international conference on applied mechanics, mechatronics and intelligent systems, pp 853–863
39. Spelko A et al (2017) CIGRE/CIRED JWG C4.42: overview of common methods for assessment of harmonic contribution from customer installation. In: 2017 IEEE Manchester Power Tech, pp 1–6
40. Chupeng X, Zejing Q, Sheng D, Chaoyang X, Zhiqi W, Yue L (2016) Effectiveness analysis of determining the main harmonic source by harmonic active power direction method. In: 2016 IEEE international conference on power and renewable energy effectiveness, pp 1–5
41. Papic I et al (2019) A benchmark test system to evaluate methods of harmonic contribution determination. IEEE Trans Power Deliv 34(1):23–31
42. Supriya P, Nambiar TNP (2012) Review of harmonic source identification techniques. Int Rev Electr Eng 7(3):4525–4531

43. Safargholi F, Malekian K, Schufft W (2018) On the dominant harmonic source identification—Part I: Review of methods. IEEE Trans Power Deliv 33(3):1268–1277
44. Safargholi F, Malekian K, Schufft W (2018) On the dominant harmonic source identification—Part II: Application and interpretation of methods. IEEE Trans Power Deliv 33(3):1278–1287
45. Bazina M, Tomiša T (2014) Comparison of various methods for determining direction of harmonic distortion by measuring in point of common coupling. In: 2014 IEEE international energy conference, pp 392–399
46. Durdhavale SR, Ahire DD (2016) A review of harmonics detection and measurement in power system. Int J Comput Appl 143(10):42–45
47. Peterson B, Rens J, Botha G, Desmet J (2015) A discriminative approach to harmonic emission assessment. In: 2015 IEEE international workshop on applied measurements for power systems, pp 7–12
48. Cataliotti A, Cosentino V, Nuccio S (2008) Comparison of nonactive powers for the detection of dominant harmonic sources in power systems. IEEE Trans Instrum Meas 57(8):1554–1561
49. Zhao J, Yang C, Xue Z, Yuan L (2017) An identification method of main harmonic current indicator based on active power direction method. In: MATEC web of conferences, vol 100, pp 1–9
50. Sinvula R, Abo-al-ez KM, Kahn MT (2019) Efficiency in distribution networks with harmonic distortion. In: 2019 Proceedings of the 17th industrial and commercial use of energy conference, pp 137–144
51. DIgSILENT GmbH (2019) DIgSILENT powerfactory technical reference documentation rectifier/inverter
52. Kundur P (1993) Power system stability and control. McGraw-Hill
53. Meegahapola L, Robinson D (2016) Dynamic modelling, simulation and control of a commercial building microgrid. In: Smart power systems and renewable energy system integration, pp 119–140
54. DIgSILENT GmbH (2019) DIgSILENT template documentation VSD template
55. XEMC Darwind. Technical data sheet. [Online]. http://www.xemc-darwind.com/Wind-turbines/XE93-2MW. Accessed 06 Mar 2020
56. Gosbell VJ, Barr RA (2014) The control of voltage THD in MV power systems. In: 2014 16th international conference on harmonics and quality of power, pp 478–482

Maximum Power Point Tracking Strategies of Grid-Connected Wind Energy Conversion Systems

Ali M. Eltamaly, Mohamed A. Mohamed, and Ahmed G. Abo-Khalil

Abstract In order to achieve maximum power point tracking (MPPT) of wind energy systems, the rotating speed of wind turbines (WTs) ought to be adjusted in the constant as indicated by wind speeds. However, fast wind speed varieties and heavy inertia bargain the MPPT control of WTs. This chapter proposes a fuzzy logic controller (FLC)-based MPPT strategy for Wind Energy Conversion Systems (WECS). The performance of the proposed MPPT strategy is analyzed mathematically and verified by simulation using MATLAB/PSIM/Simulink software. The proposed method improves the speed and accuracy of MPPT. Furthermore, the simulation results have been conducted to approve the performance of the proposed MPPT strategy, and all results have confirmed the adequacy of the proposed MPPT strategy.

Keywords Maximum power point tracking (MPPT) · Wind energy conversion systems (WECS) · Fuzzy logic controller (FLC) · Wind turbine (WT)

A. M. Eltamaly (✉)
Faculty of Engineering, Electrical Engineering Department, Mansoura University, Mansoura 35511, Egypt
e-mail: eltamaly@ksu.edu.sa

Sustainable Energy Technologies Center, King Saud University, Riyadh 11421, Saudi Arabia

K.A. CARE Energy Research and Innovation Center, Riyadh 11451, Saudi Arabia

M. A. Mohamed
Faculty of Engineering, Electrical Engineering Department, Minia University, Minia 61519, Egypt

A. G. Abo-Khalil
Department of Electrical Engineering, College of Engineering, Majmaah University, Almajmaah 11952, Saudi Arabia

Department of Electrical Engineering, College of Engineering, Assuit University, Assuit 71515, Egypt

1   Introduction

The energy of the wind has played a major role in the energy systems since time immemorial especially as a windmill in producing mechanical power [1–4]. In the last century, wind energy found application in electricity generation using wind turbines (WTs) technology [5–7]. The total installed wind power capacity of the world was estimated at 539 GW in 2017 where 52.6 added in 2017 as found in the 2017 world energy report of the world wind energy association. This installed capacity has the potential to produce 944 Terawatt-hours annually (TWh/year) and accounts for 5% of worldwide electricity consumption [8]. Annual growth rates of cumulative wind power capacity have averaged 10.8% since the end of 2016, and global capacity has increased eightfold over the past decade. Figure 1 shows the top 10 countries by nameplate wind power capacity at the end of 2017 [9–11]. For the generation of energy through the wind, WTs are used, which consist basically of blades (collectors of the kinetic energy of the wind), in addition to two axes (low and high speed), the low speed being connected to a multiplier of wind speed and the high-speed connection to the generator. The current wind systems also use a series of electronic devices aimed at controlling and acting on the energy supply of the turbines, such as power converters and "soft-starters", thus realizing the interface between the generator and the electrical network.

WTs operate at wind speed values between 5 and 25 m/s (average speed). At values below 5 m/s the turbines are unable to deliver power to the system, while for winds well above 25 m/s the turbines must not operate for safety reasons due to mechanical resistance limits. For slightly higher speed values rated, the turbines

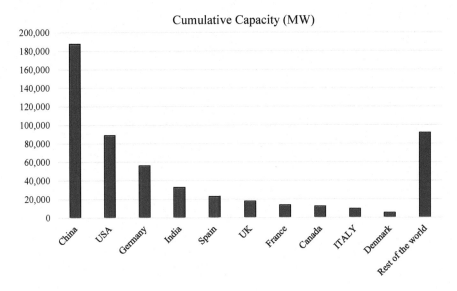

Fig. 1   Top 10 countries by nameplate wind power capacity at the end of 2017

must have control systems that reduce the use of wind energy by changing the angle of the blades in relation to the winds, thus maintaining constant power.

Typical speed values or blade tips range from 60 to 100 m/s, thus generating power between 5–10 and 2–3 MW. If connected to the grid, it is necessary to generate power with the rotor operating at a typical rotation for a frequency of 60 Hz or 50 Hz. There are two basic ways of operating wind systems:

– *Turbine operating at fixed speed*

Fixed speed wind turbines are relatively simple, consisting of a low-speed rotor rotating by the action of the winds on the turbine blades, speed multiplier, and high-speed rotor connected to the generator. These were, historically, the first to be commercially implemented.

Squirrel cage induction generators are typically used due to their simplicity. As in this type of turbine the operating speed varies very little, less than 1%, the slip of the generator varies very little too. Capacitors are usually connected in parallel with the supply circuit to correct the generator's power factor since this being an induction machine, it consumes reactive energy. "Soft-starters" must also be used to obtain a smooth growth of the magnetic flux and the electric current at the initial moment of energizing the generating unit.

– *Turbine operating at variable speed*

Nowadays it is more common to use it in large wind systems connected to the electric power networks, because over time, the size and power of the wind farms increased, thus leading to a migration of the technology of generators working at a fixed speed to variable speed. The main advantages of this technology are that it allows the connection of these generators to large electric power networks, in addition to being viable for the generation of high energy values, in addition to being used in certain cases that do not use the speed multiplier.

There are broadly two commercially existing WTs technologies namely the horizontal axis (HA-WT) and the vertical axis (VA-WT) [12–16] as detailed in the following sections. The HA-WT type has from one to three rotor blades usually directed at the wind thus the rotor has fastened tail-vanes to continuously position the blades in the path of the wind. The VA-WT type is not required to be aimed at the wind and works independently of the direction of the wind. The vertical WT however, requires more ground space because of its vertical structure [17]. WTs are often mounted on vertical structures known as towers above the ground.

1.1 Horizontal Axis WTs (HA-WTs)

This is the most renowned type of WTs. There are many designs for this type, which are shown in Fig. 2. Modern HA-WTs characteristically use two or three blades. Most European WTs are three-bladed designs [18–20]. Two-bladed designs of WTs

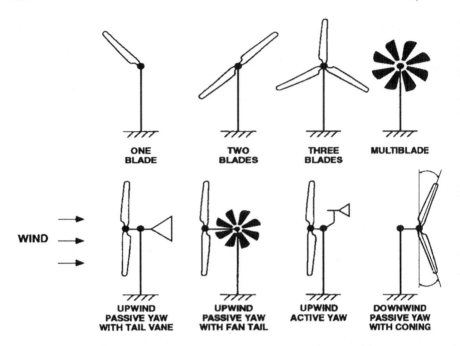

Fig. 2 HA-WT configurations

have the advantage of reducing the cost of one rotor blade and its weight. However, they require a higher rotational speed to produce the same energy output. Lately, several traditional manufacturers of two-bladed machines have switched to three-bladed machines. A one-bladed wind turbine is also possible; however, it is not very widespread commercially, due to the same problems of two-bladed design applied to one-bladed machines. In addition to higher rotational speed, they produce noise in the process of rotating the counterweight of the rotor placed on the other side of the hub from the rotor blade to balance the rotor [21–25].

Figure 3 shows the influence of the number of blades on the rotor power coefficient [26]. It is obvious from this figure that three-bladed HA-WTs have the maximum obtainable power coefficient and it works at optimal tip speed ratio. The tip speed ratio is the ratio between the tangential speed of the tip of a blade and the actual velocity of the wind [27, 28]. The noise produced by a WT is proportional to the tip speed ratio. So, the three-bladed design has lower noise which makes it the most attractive option. Also, higher tip speed ratio WTs need stronger blades to compensate for the higher centrifugal forces [29, 30].

As shown in Fig. 3, the power coefficient increases slightly with the increasing number of blades. As a result, a compromising between the increases in the power generated and the cost of extra blades should be done [32]. Most of the European manufacturers prefer three-bladed design for more stability and lesser noise while most of the USA manufacturers prefer the two-bladed design to save the cost of

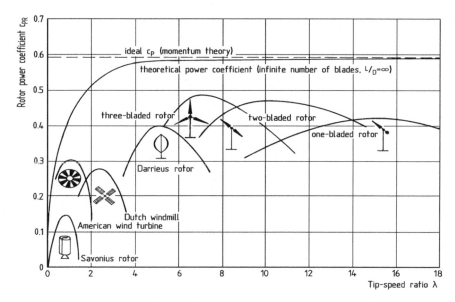

Fig. 3 Influence of the number of blades on the rotor power coefficient (envelope) and the optimum tip speed ratio [31]

one extra blade with the argument that the increase in the generation from two to three-bladed design will not compensate for the cost of the extra blade [33]. Due to the high noise associated with two-bladed design, the requirement for stronger blades, the need for high wind speed sites, and the lower power coefficient make the two-bladed design less attractive. However, the two-bladed design may be attractive also in high wind speed sites and offshore applications. The main advantages of HA-WTs are self-starting, a large variety of the rated output power (Suitable for small WTs as well as very large WTs), and comparatively low cost [34–37].

There are two main disadvantages to this type. The first is the requirement that the complete components (such as generator, gearbox, and control system) of the HA-WT be located at the top of the WT which makes maintenance difficult at such heights [38]. Secondly, when the wind changes direction, the HA-WT should be reoriented to be aligned with the change in direction [39, 40].

1. Upwind and Downwind Horizontal Axis WTs

Upwind WTs have the rotor facing the wind to make the wind hit the blades before the tower as shown in Fig. 4a [41–45]. This technique has the following features:

- Remedy the problem of spikes on the WT's output voltage due to the wind shade that the tower causes when the blades move in front of the tower especially in constant speed systems.
- It decreases the power fluctuations.

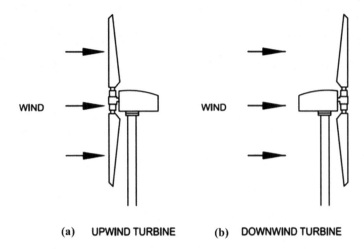

(a)   UPWIND TURBINE   (b)   DOWNWIND TURBINE

Fig. 4 Wind direction facing the wind turbine generator

- Because the blades can hit the tower it requires a rigid hub with rigid blades away from the tower to avoid touching the tower.
- This design is prominent in very large scale WTs.

Downwind WTs have the rotor on the flow-side as shown in Fig. 4b. This configuration is paramount in small and medium-size WTs. This configuration may not need a yaw mechanism if it is equipped with the streamlined body in the nacelle that will make it follow the wind [46–49]. It has the following features:

- The rotor blades are more flexible as they can bend away from the tower.
- It does not require a rigid tower as in the case of upwind.
- It suffers from the variation in output voltage and power due to the effect of the wind shadow caused by the tower.

1.2   Vertical Axis WTs (VA-WTs)

VA-WTs type is less available as compared to the HA-WTs as a result of some design constraints. In addition to the VA-WTs having a narrow range of rated output power, it also requires a starting motor and has a high comparative cost [50–52]. The main advantages of VA-WTs, are that; no additional cost is required to change the VA-WTs direction when the wind direction changes and the gearbox, generator, and control system are at the ground level thus making maintenance is very simple. Figure 5 shows different VA-WTs configurations.

Fig. 5 VA-WTs configurations [51]

1.3 Wind Resource

Wind resources are categorized into wind regimes according to the mean speeds of the sites. To install small wind turbines of less than 100 kW, annual mean wind speeds of 4.0–4.5 m/s (14.4–16.2 km/h; 9.0–10.2 mph) are required to make the system cost-effective [53]. The wind energy conversion system (WECS) most often referred to as the WT for short should be decided upon only after assessing the site wind resources. The most important data is hourly mean wind speed taken over for at least twelve months. Many years' data will provide a more accurate estimation of the annual mean wind speed of the site [54].

All nations have national meteorological administrations that record and distribute climate-related information, including wind paces and headings. The strategies are settled and composed inside the World Meteorological Association in Geneva, with the principle point of giving constant runs of information for a long time. Information have a tendency to be recorded at a moderately few for all time staffed official stations utilizing strong and confided in gear. Sadly, for wind control forecast, official estimations of wind speed tend to be estimated just at the one standard stature of 10 m, also, at stations close to airplane terminals or towns where protection from the breeze might be a characteristic component of the site. Such information are, in any case, critical as fundamental "stays" for automated breeze displaying, yet are not appropriate to apply straightforwardly to foresee wind control conditions at a particular site. Standard meteorological breeze information from the closest authority station are just valuable as first-arrange gauges; they are not adequate for definite arranging, particularly in sloping (complex) landscape. Estimations at the designated site at a few statures are expected to foresee the power created by specific turbines. Such estimations, notwithstanding for a couple of months yet best for a year, are contrasted and standard meteorological information so that the fleeting correlation might be utilized for longer term expectation; the strategy is called "measure-connect foresee". Also, data is held at expert breeze control information banks that are gotten from air ship estimations, wind control establishments and scientific demonstrating, and so forth. Such composed and open data is progressively accessible on the Internet. Wind control forecast models empower point by point wind control forecast for forthcoming breeze turbine destinations from moderately meager nearby information, even in uneven territory [54, 55].

2   Wind Energy Conversion Systems (WECS)

WT can be one of the renewable energy components of the HRES. WTs are classified from several viewpoints as discussed before. From the rotational speed perspective, there is fixed speed (FS), limited variable speed (LVS), and variable speed (VS) WTs. From the power regulation perspective WTs are classified into stall and pitch control. From the side of drive train WT is grouped into direct drive (DD) and gear drive (GD). The FS type uses a gearbox, squirrel cage induction generator (SCIG), and classified as stall, active stall, and pitch control WT. Most of the HRES are using a small WT size lower than 250 kW that may use PMSG with DC or AC output power [55, 56]. The integration configuration of the WTs depends on the output voltage type (AC or DC) as will be discussed below.

WTs can be classified according to the rotational speed concept, variable, and constant rotational speed. The variable speed operation has many advantages over constant speed operation such as increased energy capture, operation at MPPT over a wide range of wind speeds, high power quality, reduced mechanical stresses, aerodynamic noise improved system reliability, and it can provide (10–15%) higher output power and has less mechanical stresses in comparison with the operation at a fixed speed [57]. Also, WTs can be classified according to the type of drive train into direct drive (DD) and gear drive (GD). The DD operation WTs have no gearbox and have been used with small and medium-size WTs employing permanent magnet synchronous generator (PMSG) with higher numbers of poles to eliminate the need for gearbox which can be translated to higher efficiency. The GD type uses a gearbox, squirrel cage induction generator (SCIG), and classified as stall, active stall, and pitch control WT and work in constant speed applications. The variable speed WT uses doubly-fed induction generator, (DFIG) especially in high rating WTs. PMSG appears more and more attractive, because of the advantages of permanent magnet, (PM) machines over electrically excited machines such as its higher efficiency, higher energy yield, no additional power supply for the magnet field excitation, and higher reliability due to the absence of mechanical components such as slip rings. Besides, the performance of PM materials is improving, and the cost is decreasing in recent years. Therefore, these advantages make direct drive PM WT systems more attractive in the application of small and medium-scale WTs [57].

The robust controller has been developed in many literature [58–60] to track the MP available in the wind. They include tip speed ratio (TSR) [61], power signal feedback (PSF) [62], and the hill-climb searching (HCS) [59] methods. The TSR control method regulates the rotational speed of the generator to maintain an optimal TSR at which power extracted is maximum [61]. For TSR calculation, both the wind speed and turbine speed need to be measured, and the optimal TSR must be given to the controller. The first barrier to implement TSR control is the wind speed measurement, which adds to system cost and presents difficulties in practical implementations. The second barrier is the need to obtain the optimal value of TSR, this value is different from one system to another. This depends on the turbine generator characteristics results in custom-designed control software tailored for individual

WTs. In PSF control [62], it is required to have the knowledge of the wind turbine's MP curve, and track this curve through its control mechanisms. The power curves need to be obtained via simulations or offline experiments on individual WTs or from the datasheet of WT which makes it difficult to implement with accuracy in practical applications [63]. The HCS technique does not require the data of wind, generator speeds, and turbine characteristics. But, this method works well only for very small WT inertia. For large inertia WTs, the system output power is interlaced with the turbine mechanical power and rate of change in the mechanically stored energy, which often renders the HCS method ineffective [59]. On the other hand, different algorithms have been used for MP extraction from WT in addition to the three methods mentioned above. For example, reference [55] presents an algorithm for MP extraction and reactive power control of an inverter through the power angle, δ of the inverter terminal voltage, and the modulation index, m_a based variable speed WT without a wind speed sensor. Reference [64] presents an algorithm for MPPT via controlling the generator torque through q-axis current and hence controlling the generator speed with a variation of the wind speed. These techniques are used for a decoupled control of the active and reactive power from the WT through q-axis and d-axis current individually. Also, reference [65] presents a decoupled control of the active and reactive power from the WT, independently through q-axis and d-axis current but MPP operation of turbine system has been produced through regulating the input dc current of the dc/dc boost converter to follow the optimized current reference [65]. Reference [66] presents an algorithm for MPPT through directly adjusting DR of the dc/dc boost converter and modulation index of the PWM–VSC. Reference [67] presents the MPPT control algorithm based on measuring the dc-link voltage and current of the uncontrolled rectifier to attain the maximum available power from wind. Finally, references [68, 69] present MPPT control based on a fuzzy logic control (FLC). The function of FLC is to track the generator speed with the reference speed for MP extraction at variable speeds. The MPPT algorithms can be divided into two categories, the first one is MPPT algorithms for WT with wind speed sensor and the second one is MPPT algorithms without a wind speed sensor (sensorless MPPT controller). The wind speed sensor is normally used in conventional wind energy conversion systems, WECS for implementing the MPPT control algorithm. This algorithm increases cost and reduces the reliability of the WECS in addition to inaccuracies in measuring the wind speed. Therefore, some MPPT control methods estimate the wind speed; however, many of them require the knowledge of air density and mechanical parameters of the WECS [57]. Such methods require turbine generator characteristics that result in custom-design software tailored for individual WTs. Air density, on the other hand, depends upon climatic conditions and may vary considerably over various seasons. Therefore, this technique is not favorite in the modern design of WT and a lot of research efforts are focused on developing wind speed sensorless MPPT controller which does not require the knowledge of air density and turbine mechanical parameters. Therefore, the cost and maintenance of the power control system are decreased and implementation of the power control system is not difficult compared to the sensor MPPT controller [70].

According to [71], only a portion of the kinetic energy of the wind that reaches the area of the turbine blades is converted into the rotational energy of the rotor. This conversion of part of the kinetic energy of the winds causes a reduction in its speed right after it passes through the blades of the wind turbine. If we try to extract all the kinetic energy from the incident wind, the air would end at zero speed, that is, the air could not leave the turbine. In this case, we would not be able to extract energy, since obviously all the air would also be prevented from entering the turbine rotor. In the other extreme case, the wind speed after passing through the blades would remain the same as the speed before passing. With that, we would not have extracted energy from the wind either.

It is possible to assume that there must be some form of a reduction in the speed of the wind that is between these two extremes and is the most efficient situation to convert the kinetic energy of the winds into rotational energy in the rotor. According to [71], the answer is to reduce the wind at the turbine output to 2/3 of its original speed. This ideal operating point is known as Betz's Law, whose main result says that the maximum portion of the kinetic energy of the wind that can be converted into mechanical energy by a wind turbine is 59% [72].

The portion of the wind power converted into rotational energy of the rotor and which will be transmitted to the electric generator through the gear system is then defined by multiplying the power coefficient by the total wind power:

$$P_m = \frac{1}{2} C_P(\lambda, \beta)\, \rho\, A\, u^3 \tag{1}$$

where,

C_p Turbine power coefficient.
ρ Air density (kg/m^3).
A Turbine sweeping area (m^2).
u wind speed (m/s).
λ tip speed ratio of the WT which is given by the following equation [57];

$$\lambda = \frac{r_m\, \omega_r}{u} \tag{2}$$

where, r_m is the turbine rotor radius, ω_r is the angular velocity of the turbine (rad/s).

The turbine power coefficient, C_p, describes the power extraction efficiency of the WT. It is a nonlinear function of both tip speed ratio, λ, and the blade pitch angle, β. While its maximum theoretical value is approximately 0.59, it is practically between 0.4 and 0.45. There are many different versions of fitted equations for C_p made in the literature. A generic equation has been used to model $C_p(\lambda, \beta)$ and based on the modeling turbine characteristics as appeared in the following equation [57]:

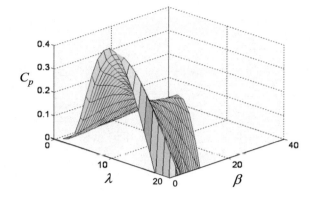

Fig. 6 Aerodynamic power coefficient variation against λ and β

$$C_P(\lambda, \beta) = 0.5176\left(116 * \frac{1}{\lambda_i} - 0.4\beta - 5\right)e^{-\frac{21}{\lambda_i}} + 0.0068\lambda \tag{3}$$

$$\text{With } \frac{1}{\lambda_i} = \frac{1}{\lambda + 0.08\beta} - \frac{0.035}{1 + \beta^3} \tag{4}$$

The C_p-λ characteristics, for different values of the pitch angle β, are illustrated in Fig. 6. The maximum value of C_p is achieved for $\beta = 0°$. The particular value of λ is defined as the optimal value (λ_{opt}). Continuous operation of WT at this point guarantees the maximum available power which can be harvested from the available wind at any speed.

2.1 Electric Generators—Variable Speed Turbines

1. Doubly-Fed Induction Generator (DFIG)

DFIG generators are of the induction type with a coiled rotor and use brushes to feed the rotor windings. The brushes receive energy through a power converter, which can decouple the operating frequency from the grid from the operating frequency of the turbine rotor. The stator is supplied directly by connecting to the grid. Usually, a circuit element called a "crowbar" is used to protect the converter. It has the function of providing a path for the overcurrent arising from electrical transients in the system when they exceed a certain level of design.

The main advantages of this type of machine are the operation with maximum efficiency, the control of active and reactive power, in addition to using lower capacity converters when compared to the FRC configuration. The main disadvantage is the need for slip rings with brushes, thus requiring constant maintenance [73–75]. A typical composition of this wind turbine configuration is shown in Fig. 7 [76].

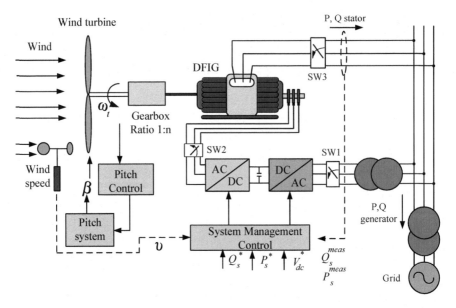

Fig. 7 Typical configuration of DFIG turbines

2. *Fully Rated Converter Wind Turbine (FRC)—Wind Turbine with Full Converter*

This type of configuration is quite versatile, as it may or may not need a speed multiplier and, besides, a range of electric generators can be used: squirrel cage and synchronous generators. As all of the turbine energy flows through converters, the dynamic operation of the generator is isolated from the grid [76].

This isolation is essential for operating at variable speed, as the operating speed of the turbine and, consequently, the operating frequency varies with the wind speed, while the network frequency is stable and varies very little under normal operating conditions. Thus, it is noticed that the power converter is the fundamental device to obtain harmonic integration between variable speed turbines with electrical networks. A typical configuration for this type of system is shown in Fig. 8 [77].

As can be seen in Fig. 8, converters associated with a dc bus can be used interconnecting the turbine with the network output transformer.

A typical operating configuration is to use the grid side converter to keep the voltage level constant, while the converter next to the generator acts to control the generator torque. An alternative way of working is to control the torque of the turbine through the converter next to the grid, while active power is transmitted from the generator to the converters [78].

This system has a disadvantage with the cost of the power converter, since now as all the energy flows through it, it will be necessary to increase the rated power of the converter and thus increase the system.

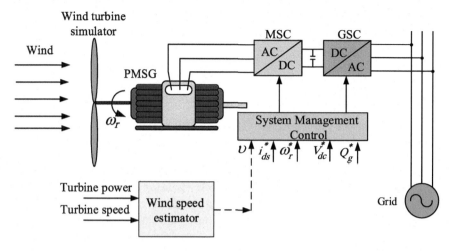

Fig. 8 Typical configuration of FRC turbines [77]

2.2 Wind Generators—Impacts on the Electrical Network

Wind turbines have their peculiarities of operation and integration with the electricity grid when compared to traditional means of power generation (hydroelectric, thermoelectric, and nuclear). Some differences in basic information can be cited [79, 80]:

- The control system for wind turbines is different, with the use of power converters being quite common;
- The driving force that generates energy, the wind, is typically a phenomenon subject to occasional variations and is therefore not controllable;
- The dimensions of the wind generator are much smaller when compared to those of conventional generators and, therefore, a large number of units make up the plant.

Thus, the behavior of the wind turbine in relation to the electrical networks is different, requiring detailed treatment.

1. **Impact on Local Networks**
– **Voltage control**

The influence of wind turbines on local energy networks depends on the type of operation chosen, that is, operation at fixed or variable speed.

For fixed speed, the generator used is the squirrel cage induction generator, which has advantages such as being more economical and easier maintenance compared to other models. However, it is not able to control the grid voltage alone. Thus, the voltage control is done by inserting reactive power through the use of elements external to the system, as in the case of capacitor banks in parallel.

Variable speed turbines are able, in principle, to vary the reactive power exchanged with the grid and, thus, influence the stabilization of the voltage level in the local power grid. This characteristic will depend on the type of converter used specifically for each turbine [81–84].

– *Protection Coordination*

As wind turbines that operate at fixed speed use induction generators, in the event of balanced faults, these only contribute to sub-transitory currents. For unbalanced faults, the contribution of wind turbines to the fault current value is integral.

For turbines operating at variable speed, DFIG generators have at first also contributed to the fault currents in the network. However, as these generators are usually associated with power converters and these converters are quite sensitive to over-currents and have to be quickly disconnected from the grid [85, 86]. Thus, by design, these turbines are quickly disconnected from the network in the event of faults, except for some caveats that are imposed by network codes for the "Ride through Fault" capability.

– *Energy Quality*

Harmonic distortions are more associated with turbines that operate at variable speed, as they use power converters that are great sources of high-frequency harmonic currents. In turbines operating at fixed speed, variations in wind speed are directly transmitted to the generator's output power, thus causing small voltage variations. If the network to which these turbines are connected is "weak", these small voltage variations will be able to generate the phenomenon known as voltage fluctuations and, consequently, flicker [87].

2. *Impact on Global Networks*
– *Grid stability*

For turbines operating at fixed speed using an induction type generator with a squirrel cage rotor, a serious problem that can occur is the rotor's over-speed. The trip occurs due to the occurrence of a fault in the system and the consequent voltage drop of that fault will generate a serious imbalance between the mechanical power generated by the wind and the power generated to the network. After the fault ceases, another problem occurs, which is the absorption of reactants from the generator through the network, contributing to delay in the recovery of voltage in the network. If the system voltage is not restored to nominal values quickly, the turbines tend to accelerate and absorb more reactive power. Thus, it is noticed that fixed speed turbines composed of squirrel cage generators are not able to help maintain the stability of the network, a fact that is essential for delivering quality energy to the consumer.

For variable speed turbines, a risk to the stability of the system is the high sensitivity of the power converters to variations in voltage and current. Thus, if the system has a large presence of this type of turbine (current market trend), and these disconnect for small and medium voltage variations, a large voltage drop will affect the entire energy system. To avoid this problem, the network codes in general establish the levels of voltage sags to which wind turbines must withstand without disconnecting, thus avoiding large losses of generation power [88, 89].

3 MPPT Control Strategies for the WECS

3.1 Power Signal Feedback (PSF) Control

In PSF control, it is required to have the knowledge of the wind turbine's MP curve, and track this curve through its control mechanisms. The MP curves need to be obtained via simulations or offline experiments on individual WTs or from the datasheet of WT which makes it difficult to implement with accuracy in practical applications. In this method, reference power is generated using an MP data curve or using the mechanical power equation of the WT where wind speed or the rotational speed is used as the input. Figure 9 shows the block diagram of a WECS with the PSF controller for MP extraction. The PSF control block generates the optimal power command P_{opt} which is then applied to the grid side converter control system for MP extraction as appeared in the following equation [72]:

$$P_{opt} = K_{opt} * \omega_r^3 \qquad (5)$$

The actual power output, P_t is compared to the optimal power, P_{opt} and any mismatch is used by the fuzzy logic controller to change the modulation index of the grid side converter, PWM-VSC as appeared in Fig. 9. The PWM-VSC is used to interface the WT with the electrical utility and will be controlled through the power angle, δ, and modulation index, m_a to control the active and reactive power output from the WTG [72].

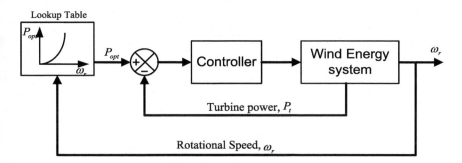

Fig. 9 The block diagram of power signal feedback control [72]

3.2 Optimal Torque Control

The torque controller aims to optimize the efficiency of wind energy capture in a wide range of wind velocities, keeping the power generated by the machine equal to the optimal defined value. It can be observed from the block diagram represented in Fig. 10, that the idea of this method is to adjust the PMSG torque according to the optimal reference torque of the WT at a given wind speed. A typical WT characteristic with the optimal torque-speed curve plotted to intersect the $C_{P\text{-max}}$ points for each wind speed is illustrated in Fig. 11. The curve T_{opt} defines the optimal torque of the device (i.e. maximum energy capture), and the control objective is to keep the turbine on this curve as the wind speed varies. For any wind speed, the MPPT device imposes a torque reference able to extract the MP. The curve T_{opt} is defined by the following equation [70]:

$$T_{opt} = K_{opt} * \omega_{opt}^2 \tag{6}$$

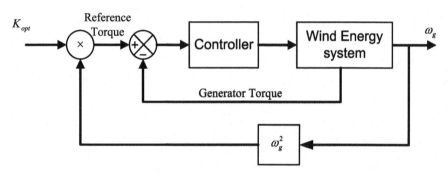

Fig. 10 The block diagram of optimal torque control MPPT method

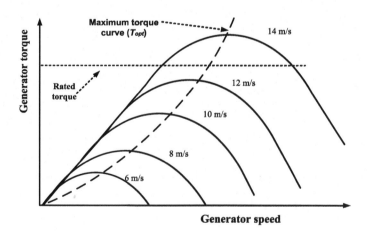

Fig. 11 Wind turbine characteristic for MP extraction [70]

where, $K_{opt} = 0.5 * \rho A * \left(\frac{r_m}{\lambda_{opt}}\right)^3 * C_{P-\max}$.

4 Decoupled Control of the Active and Reactive Power

In this study, a simple ac-dc-ac power conversion system and proposed a modular control strategy for grid-connected wind power generation systems have been implemented [65]. Grid side inverter maintains the dc-link voltage constant and the power factor of the line side can be adjusted. Input current reference of DC/DC boost converter is decided for the MPPT of the turbine without any information on wind or generator speed. As the proposed control algorithm does not require any speed sensor for wind or generator speed, construction and installation are simple, cheap, and reliable. The main circuit and control block diagrams have appeared in Fig. 12. For a wide range of variable speed operations, a dc-dc boost converter is utilized between a 3-phase diode rectifier and PWM-VSC. The input dc current is regulated to follow the optimized current reference for MPP operation of the turbine system. Grid PWM-VSC supply currents into the utility line by regulating the dc-link voltage. The active power is controlled by q-axis current through regulating the dc-link voltage whereas the reactive power can be controlled by d-axis current via adjusting the power factor of the grid side converter. The phase angle of utility voltage is detected using Phase Locked Loop, PLL, in d-q synchronous reference frame [65].

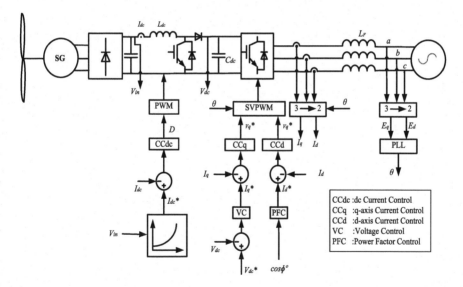

Fig. 12 Block diagram of system control [65]

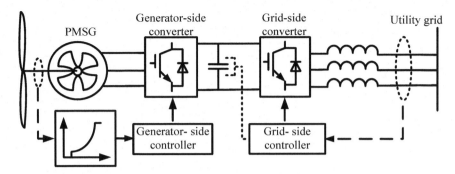

Fig. 13 Schematic diagram of the overall system [52, 55, 56]

4.1 Co-Simulation (PSIM/MATLAB) Program for Interconnecting Wind Energy System with Electric Utility

In this study, the WECS is designed as PMSG connected to the grid via a back-to-back PWM-VSC as appeared in Fig. 13. MPPT control algorithm has been introduced using FLC to regulate the rotational speed to force the PMSG to work around its MPP in speeds below rated speeds and to produce the rated power in wind speed higher than the rated wind speed of the WT. An indirect vector-controlled PMSG system has been used for this purpose. The input to FLC is two real-time measurements which are the change of output power and rotational speed between two consequent iterations (ΔP, and $\Delta \omega_m$). The output from FLC is the required change in the rotational speed $\Delta \omega_{m\text{-}new}$*. The detailed logic behind the newly proposed technique is explained in detail in the following sections. Two effective computer simulation software packages (PSIM and SIMULINK) have been used to carry out the simulation effectively where PSIM contains the power circuit of the WECS and MATLAB/SIMULINK contains the control circuit of the system. The idea behind using these two different software packages is the effective tools provided with PSIM for power circuit and the effective tools in SIMULINK for control circuit and FLC.

4.2 Wind Energy Conversion System Description

Figure 14 shows a co-simulation (PSIM/SIMULINK) program for interconnecting WECS to electric utility. The PSIM program contains the power circuit of the WECS and MATLAB/SIMULINK program contains the control of this system. The interconnection between PSIM and MATLAB/SIMULINK has been done via the SimCoupler block. The basic topology of the power circuit which has PMSG-driven WT connected to the utility grid through the ac-dc-ac conversion system has appeared in Fig. 13.

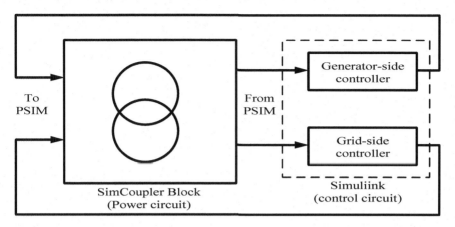

Fig. 14 Co-simulation block of wind energy system interfaced with the electric utility

The PMSG is connected to the grid through back-to-back bidirectional PWM voltage source converters VSC. The generator side converter is used as a rectifier, while the grid side converter is used as an inverter. The generator side converter is connected to the grid side converter through dc-link capacitor. The control of the overall system has been done through the generator side converter and the grid side converter. MPPT algorithm has been achieved by controlling the generator side converter using FLC. The grid side converter controller maintains the dc-link voltage at the desired value by exporting active power to the grid and it controls the reactive power exchange with the grid.

1. *PMSG Model*

The voltage equation of the three-phase PMSG type is defined as follows [73].

$$
\begin{bmatrix} v_{as} \\ v_{bs} \\ v_{cs} \end{bmatrix} = \begin{bmatrix} R_s + \frac{d}{dt}(l_{as} + M) & \frac{-1}{2}\frac{d}{dt}M & \frac{-1}{2}\frac{d}{dt}M \\ \frac{-1}{2}\frac{d}{dt}M & R_s + \frac{d}{dt}(l_{bs} + M) & \frac{-1}{2}\frac{d}{dt}M \\ \frac{-1}{2}\frac{d}{dt}M & \frac{-1}{2}\frac{d}{dt}M & R_s + \frac{d}{dt}(l_{cs} + M) \end{bmatrix} \begin{bmatrix} i_{as} \\ i_{bs} \\ i_{cs} \end{bmatrix} + \begin{bmatrix} e_{as} \\ e_{bs} \\ e_{cs} \end{bmatrix}
\tag{7}
$$

where v_{as}, v_{bs}, *and* v_{cs} are stator a, b, and c phase voltages, and i_{as}, i_{bs}, *and* i_{cs} are stator a, b, and c phase currents. The electromotive force induced in the stator windings of phases a, b, and c are e_{as}, e_{bs}, and e_{cs}, Rs is the stator winding resistance, *ls* is the leakage inductance of the stator winding, and M is the mutual inductance between the stator windings. In a balanced 3-phase circuit, the sum of each phase current becomes 0, and when $Ls = ls + (3/2)M$ (equivalent inductance (L_s)), the voltage equation is simplified as follows:

$$\begin{bmatrix} v_{as} \\ v_{bs} \\ v_{cs} \end{bmatrix} = \begin{bmatrix} R_s + \frac{d}{dt}L_s & 0 & 0 \\ 0 & R_s + \frac{d}{dt}L_s & 0 \\ 0 & 0 & R_s + \frac{d}{dt}L_s \end{bmatrix} \begin{bmatrix} i_{as} \\ i_{bs} \\ i_{cs} \end{bmatrix} + \begin{bmatrix} e_{as} \\ e_{bs} \\ e_{cs} \end{bmatrix} \qquad (8)$$

$$v_{as} = R_s i_{as} + L_s \frac{d}{dt} i_{as} + e_{as} \qquad (9)$$

$$v_{bs} = R_s i_{bs} + L_s \frac{d}{dt} i_{bs} + e_{bs} \qquad (10)$$

$$v_{cs} = R_s i_{cs} + L_s \frac{d}{dt} i_{cs} + e_{cs} \qquad (11)$$

Equations (9)–(11) can be integrated into a vector form as follows:

$$v_{abcs} = R_s I_{abcs} + L_s \frac{d I_{abcs}}{dt} + E_{abcs} \qquad (12)$$

If the number of fluxes crossing the stator winding by the rotor of the PMSG is called *epsa*, *epsb*, and *epsc*, and the maximum value of the flux is *epsf*, the magnitude of the flux crossing the stator winding is as follows:

$$\Psi_a = \Psi_f \cos \theta$$

$$\Psi_b = \Psi_f \cos \left(\theta - \frac{2}{3}\pi \right)$$

$$\Psi_c = \Psi_f \cos \left(\theta + \frac{2}{3}\pi \right) \qquad (13)$$

At this time, the synchronous phase angle θ was calculated as the rotational speed ω_r. If $\theta = \int \omega_r dt$, the electromotive force induced in the three-phase stator winding is defined as follows:

$$e_{as} = -\omega_r \Psi_f \sin\theta$$

$$e_{bs} = -\omega_r \Psi_f \sin \left(\theta - \frac{2}{3}\pi \right)$$

$$e_{cs} = -\omega_r \Psi_f \sin \left(\theta + \frac{2}{3}\pi \right) \qquad (14)$$

To convert a 3-phase state equation to a *d-q* stationary reference frame ($\omega = 0$), multiply the following transformation matrix:

$$T(0) = \begin{bmatrix} 1 & 0 & 0 \\ 0 & \frac{1}{\sqrt{3}} & \frac{-1}{\sqrt{3}} \end{bmatrix} \tag{15}$$

Multiplying Eq. (12) by Eq. (15):

$$T(0)v_{abcs} = R_s T(0)I_{abcs} + L_s \frac{dI_{abcs}}{dt}T(0) + T(0)E_{abcs}$$

$$V_{dqs}^S = R_s I_{dqs}^S + L_s \frac{d}{dt}I_{dqs}^S + E_{dqs}^S \tag{16}$$

The electromotive force induced to the stator is obtained as follows by substituting Eq. (13) into Eq. (16) as follows:

$$E_{abcs} = \begin{bmatrix} -\omega_r \Psi_f \sin\theta \\ \omega_r \Psi_f \cos\theta \end{bmatrix} \tag{17}$$

The flux linkage of the stationary system is as follows:

$$\begin{bmatrix} \Psi_{ds}^s \\ \Psi_{qs}^s \end{bmatrix} = T(0) \begin{bmatrix} \Psi_f \cos\theta \\ \Psi_f \cos(\theta - \frac{2}{3}\pi) \\ \Psi_f \cos(\theta + \frac{2}{3}\pi) \end{bmatrix} = \begin{bmatrix} \Psi_f \cos\theta \\ \Psi_f \sin\theta \end{bmatrix} \tag{18}$$

To perform the control of a rotating device, the state equation of the stationary coordinate system must be converted into a synchronous coordinate system:

$$V_{dqs}^S = R_s I_{dqs}^e + L_s \frac{d}{dt}I_{dqs}^e + L_s \begin{bmatrix} 0 & -\omega_r \\ -\omega_r & 0 \end{bmatrix} + E_{dqs}^e \tag{19}$$

If the 3-phase state equation is changed to the synchronous coordinate system d-q, it is as shown in Eq. (19), and rearranged as follows:

$$\begin{bmatrix} v_{ds}^e \\ v_{qs}^e \end{bmatrix} = \begin{bmatrix} R_s + \frac{dL_s}{dt} & -\omega_r L_s \\ -\omega_r L_s & R_s + \frac{dL_s}{dt} \end{bmatrix} \begin{bmatrix} i_{ds}^e \\ i_{qs}^e \end{bmatrix} + \begin{bmatrix} e_{ds}^e \\ e_{qs}^e \end{bmatrix} \tag{20}$$

The stator induced EMF can be expressed as follows:

$$\begin{bmatrix} e_{ds}^e \\ e_{qs}^e \end{bmatrix} = \begin{bmatrix} 0 \\ \omega_r \Psi_f \end{bmatrix} \tag{21}$$

Finally, the PMSG's synchronous coordinate system d-q voltage equation can be obtained.

$$\begin{bmatrix} v_{ds}^e \\ v_{qs}^e \end{bmatrix} = \begin{bmatrix} R_s + \frac{dL_s}{dt} & -\omega_r L_s \\ -\omega_r L_s & R_s + \frac{dL_s}{dt} \end{bmatrix} \begin{bmatrix} i_{ds}^e \\ i_{qs}^e \end{bmatrix} + \begin{bmatrix} 0 \\ \omega_r \Psi_f \end{bmatrix} \tag{22}$$

Since the rotor of the PMSG has no energy input and output, it can be regarded as electromagnetic torque, which can be expressed as follows:

$$T_e = \frac{3}{2} P \psi_F i_{qs} \tag{23}$$

where, P is the number of pole pairs.

5   Control of the Generator Side Converter (PMSG)

The generator side controller controls the rotational speed to produce the maximum output power via controlling the electromagnetic torque according to Eq. (10), where the indirect vector control is used. The proposed control logic of the generator side converter appears in Fig. 15. The speed loop will generate the q-axis current component to control the generator torque and speed at different wind speed via estimating the reference value of i_α, i_β as appeared in Fig. 15. The torque control can be achieved through the control of the i_{sq} current. Figure 16 shows the stator and rotor current space phasors and the excitation flux of the PMSG. The quadrature stator current i_{sq} can be controlled through the rotor reference frame (α, β axis) as appeared in Fig. 16. So, the reference value of i_α, i_β can be estimated easily from the amplitude of i_{sq*} and the rotor angle, θ_r. Initially, to find the rotor angle, θ_r, the relationship between the electrical angular speed, ω_r, and the rotor mechanical speed (rad/s), ω_m may be expressed as:

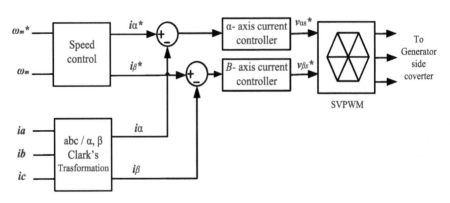

Fig. 15 Control block diagram of PMSG

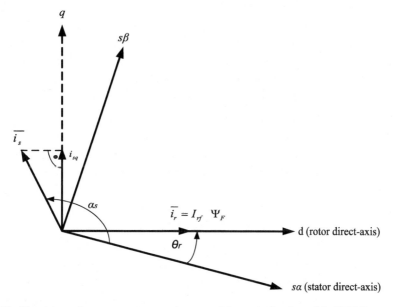

Fig. 16 The stator and rotor current space phasors and the excitation flux of the PMSG

$$\omega_r = \frac{P}{2}\omega_m \tag{24}$$

So, the rotor angle, θ_r, can be estimated by integrating the electrical angular speed, ω_r. The input to the speed control is the actual and reference rotor mechanical speed (rad/s) and the output is the (α, β) reference current components. The actual values of the (α, β) current components are estimated using Clark's transformation to the three-phase current of PMSG. The FLC can be used to find the reference speed along which tracks the MPP.

5.1 Fuzzy Logic Controller for MPPT

At a certain wind speed, the power is maximum at a certain ω called optimum rotational speed, ω_{opt}. This speed corresponds to the optimum tip speed ratio, λ_{opt} [71]. So, to extract MP at variable wind speed, the turbine should always operate at λ_{opt}. This occurs by controlling the rotational speed of the turbine. Controlling the turbine to operate at optimum rotational speed can be done using the fuzzy logic controller. Each WT has one value of λ_{opt} at variable speed but ω_{opt} changes from certain wind speed to another. From Eq. (6), the relation between ω_{opt} and wind speed, u, for constant R and λ_{opt} can be deduced as appeared in (25):

$$\omega_{opt} = \frac{\lambda_{opt}}{R}u \tag{25}$$

From Eq. (25), the relation between the optimum rotational speed and wind speed is linear. At a certain wind speed, there is an optimum rotational speed which is different at another wind speed. The fuzzy logic control is used to search (observation and perturbation) the rotational speed reference which tracks the MPP at variable wind speeds. The fuzzy logic controller block diagram appears in Fig. 17. Two real-time measurements are used as input to fuzzy (ΔP, and $\Delta\omega_m*$) and the output is ($\Delta\omega_{m\text{-}new}*$). Membership functions appear in Fig. 18. Triangular symmetrical membership functions are suitable for the input and output, which give more sensitivity especially as variables approach zero value. The width of variation can

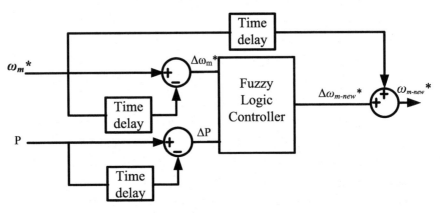

Fig. 17 Input and output of fuzzy controller

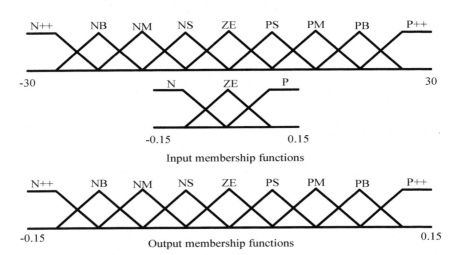

Fig. 18 Membership functions of the fuzzy logic controller

Table 1 Rules of fuzzy logic controller

$\frac{\Delta P}{\Delta \omega_m}$	N++	NB	NM	NS	ZE	PS	PM	PB	P++
N	P++	PB	PM	PS	ZE	NS	NM	NB	N++
ZE	NB	NM	NS	NS	ZE	PS	PM	PM	PB
P	N++	NB	NM	NS	ZE	PM	PM	PB	PB

be adjusted according to the system parameter. The input signals are first fuzzified and expressed in fuzzy set notation using linguistic labels which are characterized by membership functions before it is processed by the FLC. Using a set of rules and a fuzzy set theory, the output of the FLC is obtained. This output, expressed as a fuzzy set using linguistic labels characterized by membership functions, is defuzzified and then produces the controller output. The fuzzy logic controller doesn't require any detailed mathematical model of the system and its operation is governed simply by a set of rules. The principle of the fuzzy logic controller is to perturb the reference speed ω_m* and to observe the corresponding change of power, ΔP. If the output power increases with the last increment, the searching process continues in the same direction. On the other hand, if the speed increment reduces the output power, the direction of the searching is reversed. The fuzzy logic controller is efficient to track the MPP, especially in case of frequently changing wind conditions.

Figure 18 shows the input and output membership functions and Table 1 lists the control rule for the input and output variable. The next fuzzy levels are chosen for controlling the inputs and output of the fuzzy logic controller. The variation step of the power and the reference speed may vary depending on the system. In Fig. 18, the variation step in the speed reference is from −0.15 to 0.15 rad/s for power variation ranging over from −30 to 30 W. The membership definitions are given as follows: N (negative), N++ (very big negative), NB (negative big), NM (negative medium), NS (negative small), ZE (zero), P (positive), PS (positive small), PM (positive medium), PB (positive big), and P++ (very big positive).

5.2 Control of the Grid Side Converter

The power flow of the grid side converter is controlled in order to maintain the dc-link voltage at reference value, 600 V. Since increasing the output power than the input power to dc-link capacitor causes a decrease of the dc-link voltage and vice versa, the output power will be regulated to keep dc-link voltage approximately constant. To maintain the dc-link voltage constant and to ensure the reactive power flowing into the grid, the grid side converter currents are controlled using the d-q vector control approach. The dc-link voltage is controlled to the desired value by using a PI-controller and the change in the dc-link voltage represents a change in the q-axis (i_{qs}) current component. Figure 19 shows a control block diagram of the grid side converter. The active power can be defined as appeared in the following equation:

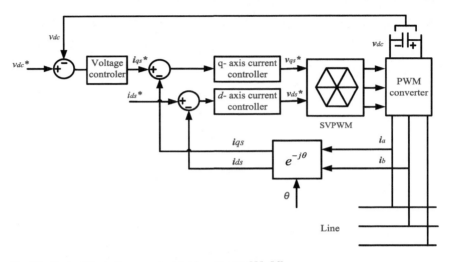

Fig. 19 Control block diagram of grid side converter [55, 56]

$$P_s = \frac{3}{2}\left(v_{ds}i_{ds} + v_{qs}i_{qs}\right) \tag{26}$$

The reactive power can be defined as:

$$Q_s = \frac{3}{2}\left(v_{qs}i_{ds} - v_{ds}i_{qs}\right) \tag{27}$$

By aligning the q-axis of the reference frame along with the grid voltage position $v_{ds} = 0$ and $v_{qs} = $ constant because the grid voltage is assumed to be constant. Then the active and reactive power can be obtained from the following equations:

$$P_s = \frac{3}{2}v_{qs}i_{qs} \tag{28}$$

$$Q_s = \frac{3}{2}v_{qs}i_{ds} \tag{29}$$

6 Simulation Results and Discussion

A co-simulation (PSIM/SIMULINK) program has been used where PSIM contains the power circuit of the WECS and MATLAB/SIMULINK has the whole control system as described before. The model of WECS in PSIM contains the WT connected to the utility grid through a back-to-back bidirectional PWM converter. The control of the whole system in SIMULINK contains the generator side controller and the grid side controller. The generator can be directly controlled by the generator side controller to track the MP available from the WT. The wind speed is variable and changes from 7 to 13 m/s as input to WT. To extract MP at variable wind speed, the turbine should always operate at λ_{opt}. This occurs by controlling the rotational speed of the WT. So, it always operates at the optimum rotational speed. ω_{opt} changes from certain wind speed to another. The fuzzy logic controller is used to search the optimum rotational speed which tracks the MPP at variable wind speeds. Figure 20a shows the variation of the wind speed which varies randomly from 7 to 13 m/s. On the other hand, Fig. 20b shows the variation of the actual and reference rotational speed as a result of the wind speed variation. At a certain wind speed, the actual and reference rotational speed has been estimated and this agreement with the power characteristic of the WT appeared later in Fig. 14. i.e. the WT always operates at the optimum rotational speed which is found using FLC; hence, the power extraction from wind is maximum at variable wind speed. It is seen that according to the wind speed variation the generator speed varies and that its output power is produced corresponding to the wind speed variation. The fuzzy logic controller works well and it gives a good tracking performance for the MPP. The fuzzy logic controller makes WT always operates at the optimum rotational speed. On the other hand, the grid side controller maintains the dc-link voltage at the desired value, 600v, as appeared in Fig. 20c. The dc-link voltage is regulated by exporting active power to the grid as appeared in Fig. 20d. The reactive power transmitted to the grid appears in Fig. 20e.

7 Conclusion

Variable speed operation and direct drive wind turbines (WTs) have been considered as the modern aspects of wind energy conversion systems (WECS). In this chapter, a fuzzy logic controller (FLC)-based MPPT strategy for (WECS) is proposed. The generator side controller has been used to track the maximum power generated from WTs through controlling the rotational speed of the WT using FLC. The performance of the proposed MPPT strategy is analyzed mathematically and verified by simulation using MATLAB/PSIM/Simulink software. The PMSG has been controlled in an indirect vector field-oriented control technique and its speed reference has been obtained from FLC. The simulation outcomes have been conducted to approve the performance of the proposed MPPT strategy, and all results have confirmed the adequacy of the proposed MPPT strategy.

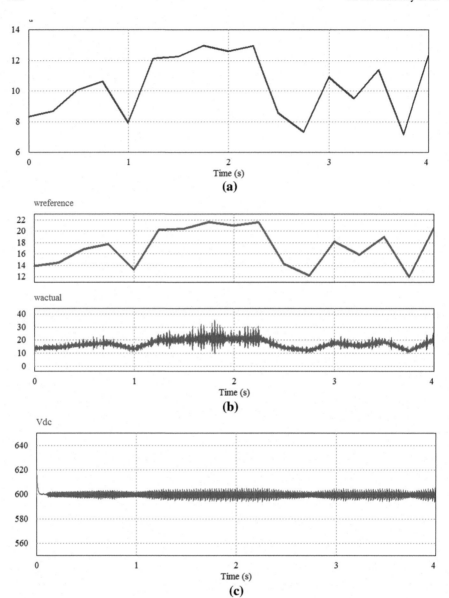

Fig. 20 Different simulation waveforms: **a** wind speed variation (7–13) m/s. **b** actual and reference rotational speed (rad/s). **c** dc-link voltage (V). **d** Active power (W). **e** Reactive power (Var)

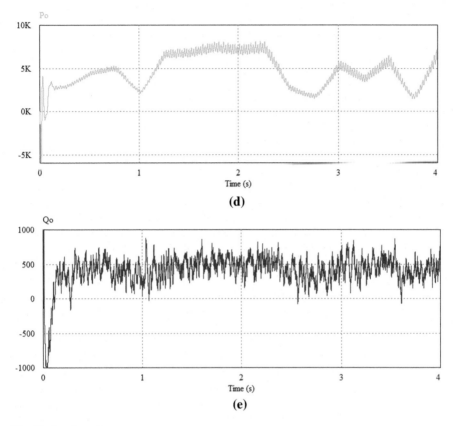

Fig. 20 (continued)

References

1. Ahmed MA, Eltamaly AM, Alotaibi MA, Alolah AI, Kim Y-C (2020) Wireless network architecture for cyber physical wind energy system. IEEE Access 8:40180–40197
2. Eltamaly AM, Addoweesh KE, Bawa U, Mohamed MA (2014) Economic modeling of hybrid renewable energy system: a case study in Saudi Arabia. Arab J Sci Eng 39(5):3827–3839
3. Eltamaly AM, Al-Saud M, Sayed K, Abo-Khalil AG (2020) Sensorless active and reactive control for DFIG wind turbines using opposition-based learning technique. Sustainability 12(9):3583
4. Eltamaly AM, Mohamed MA (2014) A novel design and optimization software for autonomous PV/wind/battery hybrid power systems. Math Probl Eng
5. Mohamed MA, Jin T, Su W (2020) An effective stochastic framework for smart coordinated operation of wind park and energy storage unit. Appl Energy 272:115228
6. Al-Saud MS, Eltamaly AM, Al-Ahmari A (2017) Multi-rotor vertical axis wind turbine. U.S. patent 9,752,556, 5 Sept 2017
7. Abo-Khalil AG, Ali Alghamdi IT, Eltamaly AM (2019) Current controller design for DFIG-based wind turbines using state feedback control. IET Renew Power Gener 13(11):1938–1948

8. Eltamaly AM, Addoweesh KE, Bawah U, Mohamed MA (2013) New software for hybrid renewable energy assessment for ten locations in Saudi Arabia. J Renew Sustain Energy 5(3):033126

9. Eltamaly AM (2014) Pairing between sites and wind turbines for Saudi Arabia Sites. Arab J Sci Eng 39(8):6225–6233

10. Jin T, Guo J, Mohamed MA, Wang M (2019) A novel model predictive control via optimized vector selection method for common-mode voltage reduction of three-phase inverters. IEEE Access 7:95351–95363

11. Eltamaly AM (2013) Design and implementation of wind energy system in Saudi Arabia. Renew Energy 60:42–52

12. Eltamaly AM (2013) Design and simulation of wind energy system in Saudi Arabia. In: 2013 4th international conference on intelligent systems, modelling and simulation. IEEE, pp 376–383

13. Eltamaly AM, Alolah AI, Farh HM, Arman H (2013) Maximum power extraction from utility-interfaced wind turbines. New Dev Renew Energy 159–192

14. Eltamaly AM, Al-Saud MS (2018) Nested multi-objective PSO for optimal allocation and sizing of renewable energy distributed generation. J Renew Sustain Energy 10(3):035302

15. Aziz MS, Mufti GM, Ahmad S (2017) Wind-hybrid power generation systems using renewable energy sources-a review. Int J Renew Energy Res (IJRER) 7(1):111–127

16. Mohamed MA, Eltamaly AM, Alolah AI, Hatata AY (2019) A novel framework-based cuckoo search algorithm for sizing and optimization of grid-independent hybrid renewable energy systems. Int J Green Energy 16(1):86–100

17. Savino MM, Manzini R, Selva VD, Accorsi R (2017) A new model for environmental and economic evaluation of renewable energy systems: the case of wind turbines. Appl Energy 189:739–752

18. Eltamaly AM, Mohamed MA (2018) Optimal sizing and designing of hybrid renewable energy systems in smart grid applications. Advances in renewable energies and power technologies. Elsevier, pp 231–313

19. Olatayo KI, Wichers JH, Stoker PW (2018) Energy and economic performance of small wind energy systems under different climatic conditions of South Africa. Renew Sustain Energy Rev 98:376–392

20. Mohamed MA, Eltamaly AM, Farh HM, Alolah AI (2015) Energy management and renewable energy integration in smart grid system. In: 2015 IEEE international conference on smart energy grid engineering (SEGE). IEEE, pp 1–6

21. Mohamed MA, Eltamaly AM, Alolah AI (2016) PSO-based smart grid application for sizing and optimization of hybrid renewable energy systems. PLoS ONE 11(8):e0159702

22. Mabrouk IB, Hami AE, Walha L, Zghal B, Haddar M (2017) Dynamic vibrations in wind energy systems: application to vertical axis wind turbine. Mech Syst Signal Process 85:396–414

23. Gandomkar A, Parastar A, Seok J-K (2016) High-power multilevel step-up DC/DC converter for offshore wind energy systems. IEEE Trans Ind Electron 63(12):7574–7585

24. Eltamaly AM, Mohamed MA, Alolah AI (2016) A novel smart grid theory for optimal sizing of hybrid renewable energy systems. Sol Energy 124:26–38

25. Liu S-Y, Ho Y-F (2016) Wind energy applications for Taiwan buildings: what are the challenges and strategies for small wind energy systems exploitation? Renew Sustain Energy Rev 59:39–55

26. Eltamaly AM, Mohamed MA (2016) A novel software for design and optimization of hybrid power systems. J Braz Soc Mech Sci Eng 38(4):1299–1315

27. Salari ME, Coleman J, Toal D (2019) Analysis of direct interconnection technique for offshore airborne wind energy systems under normal and fault conditions. Renew Energy 131:284–296

28. Eltamaly AM, El-Tamaly H, Enjeti P (2001) New modular wind energy conversion system. In: The 8th international middle-east power systems conference MEPCON' 2001, University of Helwan, Cairo, Egypt, 29–31 Dec 2001

29. Kececioglu OF, Acikgoz H, Yildiz C, Gani A, Sekkeli M (2017) Power quality improvement using hybrid passive filter configuration for wind energy systems. J Electr Eng Technol 12(1):207–216

30. Mohamed MA, Eltamaly AM (2018) A novel smart grid application for optimal sizing of hybrid renewable energy systems. Modeling and simulation of smart grid integrated with hybrid renewable energy systems. Springer, Cham, pp 39–51
31. Manwell JF, McGowan JG, Rogers AL (2010) Wind energy explained: theory, design and application. Wiley
32. Raju AB, Fernandes BG, Chatterjee K (2004) A UPF power conditioner with maximum power point tracker for grid connected variable speed wind energy conversion system. In: Proceedings. 2004 first international conference on power electronics systems and applications. IEEE, pp 107–112
33. Mohamed MA, Eltamaly AM (2018) Modeling of hybrid renewable energy system. Modeling and simulation of smart grid integrated with hybrid renewable energy systems. Springer, Cham, pp 11–21
34. Mohamed MA, Eltamaly AM, Alolah AI (2015) Sizing and techno-economic analysis of stand-alone hybrid photovoltaic/wind/diesel/battery power generation systems. J Renew Sustain Energy 7(6):063128
35. Bottiglione F, Mantriota G, Valle M (2018) Power-split hydrostatic transmissions for wind energy systems. Energies 11(12):3369
36. Mohamed MA, Eltamaly AM (2018) A PSO-based smart grid application for optimum sizing of hybrid renewable energy systems. Modeling and simulation of smart grid integrated with hybrid renewable energy systems. Springer, Cham, pp 53–60
37. El-Tamaly AM, Enjeti PN, El-Tamaly HH (2000) An improved approach to reduce harmonics in the utility interface of wind, photovoltaic and fuel cell power systems. In APEC 2000, fifteenth annual IEEE applied power electronics conference and exposition (Cat. No. 00CH37058), vol 2. IEEE, pp 1059–1065
38. Abo-Khalil AG, Lee DC (22008) Maximum power point tracking based on sensorless wind speed using support vector regression. IEEE Trans Ind Electron 55(3)
39. Mohamed MA, Eltamaly AM (2018) Sizing and techno-economic analysis of stand-alone hybrid photovoltaic/wind/diesel/battery energy systems. Modeling and simulation of smart grid integrated with hybrid renewable energy systems. Springer, Cham, pp 23–38
40. Bafandeh A, Vermillion C (2016) Real-time altitude optimization of airborne wind energy systems using Lyapunov-based switched extremum seeking control. In: 2016 American control conference (ACC). IEEE, pp 4990–4995
41. El-Tamaly AM, El-Tamaly HH, Cengelci E, Enjeti PN, Muljadi E (1999) Low cost PWM converter for utility interface of variable speed wind turbine generators. In: Fourteenth annual applied power electronics conference and exposition, APEC'99, conference proceedings (Cat. No. 99CH36285), vol 2. IEEE, pp 889–895
42. Eltamaly AM, Al-Saud MS, Abo-Khalil AG (2020) Dynamic control of a DFIG wind power generation system to mitigate unbalanced grid voltage. IEEE Access 8:39091–39103
43. Li Y, Feng Bo, Li G, Qi J, Zhao D, Yunfei Mu (2018) Optimal distributed generation planning in active distribution networks considering integration of energy storage. Appl Energy 210:1073–1081
44. Eltamaly AM (2002) New formula to determine the minimum capacitance required for self-excited induction generator. In: 2002 IEEE 33rd annual IEEE power electronics specialists conference. Proceedings (Cat. No. 02CH37289), vol 1. IEEE, pp. 106–110
45. Koutsoukis NC, Siagkas DO, Georgilakis PS, Hatziargyriou ND (2016) Online reconfiguration of active distribution networks for maximum integration of distributed generation. IEEE Trans Autom Sci Eng 14(2):437–448
46. Hamouda RM, Eltamaly AM, Alolah AI (2008) Transient performance of an isolated induction generator under different loading conditions. In: 2008 Australasian Universities power engineering conference. IEEE, pp 1–5
47. Novakovic B, Nasiri A (2017) Modular multilevel converter for wind energy storage applications. IEEE Trans Ind Electron 64(11):8867–8876
48. Eltamaly AM, Mohamed YS, El-Sayed A-HM, Elghaffar ANA (2019) Analyzing of wind distributed generation configuration in active distribution network. In: 2019 8th international conference on modeling simulation and applied optimization (ICMSAO). IEEE, pp 1–5

49. Allik A, Märss M, Uiga J, Annuk A (2016) Optimization of the inverter size for grid-connected residential wind energy systems with peak shaving. Renew Energy 99:1116–1125
50. Bawah U, Addoweesh KE, Eltamaly AM (2013) Comparative study of economic viability of rural electrification using renewable energy resources versus diesel generator option in Saudi Arabia. J Renew Sustain Energy 5(4):042701
51. Farh HM, Eltamaly AM (2013) Fuzzy logic control of wind energy conversion system. J Renew Sustain Energy 5(2):023125
52. Eltamaly AM, Farh HM (2013) Maximum power extraction from wind energy system based on fuzzy logic control. Electr Power Syst Res 97:144–150
53. Alhmoud L, Wang B (2018) A review of the state-of-the-art in wind-energy reliability analysis. Renew Sustain Energy Rev 81:1643–1651
54. Bawah U, Addoweesh KE, Eltamaly AM (2012) Economic modeling of site-specific optimum wind turbine for electrification studies. Adv Mater Res 347:1973–1986. Trans Tech Publications Ltd.
55. Eltamaly AM, Farh HM (2015) Smart maximum power extraction for wind energy systems. In: 2015 IEEE international conference on smart energy grid engineering (SEGE). IEEE, pp 1–6
56. Eltamaly AM, Alolah AI, Farh HM, Arman H (2013) Maximum power extraction from utility-interfaced wind turbines. In: New developments in renewable energy. Intech, pp 159–192
57. Tran CH, Nollet F, Essounbouli N, Hamzaoui A (2018) Maximum power point tracking techniques for wind energy systems using three levels boost converter. In: IOP conference series: earth and environmental science, vol 154, no 1. IOP Publishing, p 012016
58. Al-Shamma'a A, Addoweesh KE, Eltamaly A (2012) Optimum wind turbine site matching for three locations in Saudi Arabia. Adv Mater Res 347:2130–2139. Trans Tech Publications Ltd.
59. Wang Q, Chang L (2004) An intelligent maximum power extraction algorithm for inverter-based variable speed wind turbine systems. IEEE Trans Power Electron 19(5):1242–1249
60. Eltamaly AM, Khan AA (2011) Investigation of DC link capacitor failures in DFIG based wind energy conversion system. Trends Electr Eng 1(1):12–21
61. Li H, Shi KL, McLaren PG (2005) Neural-network-based sensorless maximum wind energy capture with compensated power coefficient. IEEE Trans Ind Appl 41(6):1548–1556
62. Al-Saud M, Eltamaly AM, Mohamed MA, Kavousi-Fard A (2019) An intelligent data-driven model to secure intravehicle communications based on machine learning. IEEE Trans Ind Electron 67(6):5112–5119
63. Eltamaly AM, Alolah AI, Abdel-Rahman MH (2011) Improved simulation strategy for DFIG in wind energy applications. Int Rev Model Simul 4(2)
64. Abir A, Mehdi D (2017) Control of permanent-magnet generators applied to variable-speed wind-energy. In: 2017 international conference on green energy conversion systems (GECS). IEEE, pp 1–6
65. Song S-H, Kang S, Hahm N (2003) Implementation and control of grid connected AC-DC-AC power converter for variable speed wind energy conversion system. In: Eighteenth annual IEEE applied power electronics conference and exposition, APEC'03, vol 1. IEEE, pp 154–158
66. Eltamaly AM (2007) Modeling of wind turbine driving permanent magnet generator with maximum power point tracking system. J King Saud Univ Eng Sci 19(2):223–236
67. Eltamaly AM, Alolah AI, Abdel-Rahman MH (2010) Modified DFIG control strategy for wind energy applications. In: Proceedings of SPEEDAM 2010. IEEE, pp 653–658
68. Mohamed MA, Eltamaly AM, Alolah AI (2017) Swarm intelligence-based optimization of grid-dependent hybrid renewable energy systems. Renew Sustain Energy Rev 77:515–524
69. Yao X, Guo C, Xing Z, Li Y, Liu S (2009) Variable speed wind turbine maximum power extraction based on fuzzy logic control. In: 2009 international conference on intelligent human-machine systems and cybernetics, vol 2. IEEE, pp 202–205
70. Eltamaly AM, Mohamed MA, Al-Saud MS, Alolah AI (2017) Load management as a smart grid concept for sizing and designing of hybrid renewable energy systems. Eng Optim 49(10):1813–1828

71. Abo-Khalil AG, Lee DC, Seok JK (2004) Variable speed wind power generation system based on fuzzy logic control for maximum output power tracking. In: Proceedings of power electronics specialists conference '04, pp 2039–2043

72. Farh HM, Eltamaly AM, Mohamed MA (2012) Wind energy assessment for five locations in Saudi Arabia. In: International conference on future environment and energy (ICFEE 2012), IPCBEE (2012) © (2012). IACSIT Press, 26–28 Feb 2012, Singapore

73. Etamaly AM, Mohamed MA, Alolah AI (2015) A smart technique for optimization and simulation of hybrid photovoltaic/wind/diesel/battery energy systems. In: 2015 IEEE international conference on smart energy grid engineering (SEGE). IEEE, pp 1–6

74. Abo-Khalil AG, Alghamdi AS, Eltamaly AM, Al-Saud MS, Praveen PR, Sayed K (2019) Design of state feedback current controller for fast synchronization of DFIG in wind power generation systems. Energies 12(12):2427

75. Alsalloum A, Hamouda RM, Alolah AI, Eltamaly AM (2010) Transient performance of an isolated induction generator under unbalanced loading conditions. J Energy Power Eng 4(5)

76. Abokhalil AG (2019) Grid connection control of DFIG for variable speed wind turbines under turbulent conditions. Int J Renew Energy Res (IJRER) 9(3):1260–1271

77. Abo-KhalilAG (2012) Synchronization of DFIG output voltage to utility grid in wind power system. Elsevier J Renew Energy 44:193–198. Impact Factor: 4.37. Abo-Khalil AG, Lee DC (2008) DC-link capacitance estimation in AC/DC/ACPWM converters using voltage injection. IEEE Trans Ind Appl 44(5). Impact Factor: 3.347

78. Abo-Khalil AG et al (2016) A low-cost PMSG topology and control strategy for small-scale wind power generation systems. Int J Eng Sci Res Technol IJESRT 585–592

79. Khaled U, Eltamaly AM, Beroual A (2017) Optimal power flow using particle swarm optimization of renewable hybrid distributed generation. Energies 10(7):1013

80. Abo-KhalilAG (2013) Impacts of wind farms on power system stability. Wind farm. ISBN 980-953-307-562-9

81. El-Tamaly HH, El-Tamaly AM, El-Baset Mohammed AA (2003) Design and control strategy of utility interfaced PV/WTG hybrid system. In: The ninth international middle east power system conference, MEPCON, pp 16–18

82. Abo-Khalil AG (2011) A new wind turbine simulator using a squirrel-cage motor for wind power generation systems. In: The power electronics and drive systems conference PEDS, Dec 2011, Singapore

83. Eltamaly AM (2012) A novel harmonic reduction technique for controlled converter by third harmonic current injection. Electr Power Syst Res 91:104–112

84. Park HG, Abo-Khalil AG, Lee DC, Son KM (2007) Torque ripple elimination for doubly-fed induction motors under unbalanced source voltage. In: Proceedings of power electronics and drive systems PEDS'07, Nov 2007, pp 1301–1306

85. Eltamaly AM, Al-Shamma'a AA (2016) Optimal configuration for isolated hybrid renewable energy systems. J Renew Sustain Energy 8(4):045502

86. Abo-Khalil AG, Lee DC (2006) Dynamic modeling and control of wind turbines for grid-connected wind generation system. In: Proceedings on power electronics specialists conference, Korea

87. Eltamaly AM (2009) Harmonics reduction techniques in renewable energy interfacing converters. Renew Energy. Intechweb

88. Abo-Khalil AG, Lee DC (2006) Grid connection of doubly-fed induction generators in wind energy conversion. In: Proceedings of IPEMC, China, Aug 2006

89. Eltamaly AM (2000) Power quality consideration for interconnecting renewable energy power converter systems to electric utility. PhD diss, PhD thesis, Elminia University

Hybrid CSA-GWO Algorithm-Based Optimal Control Strategy for Efficient Operation of Variable-Speed Wind Generators

Mina N. Amin, Mahmoud A. Soliman, Hany M. Hasanien, and Almoataz Y. Abdelaziz

Abstract The hybridization of two algorithms is recently emerging to find superior solutions to the optimization problems. This paper exhibits a new hybrid cuckoo search algorithm and grey wolf optimizer (CSA-GWO)-based on control scheme to enhance the transient stability of the grid-tied variable-speed wind energy conversion system. The variable-speed wind turbine (VSWT) direct drive permanent-magnet synchronous generator, which is interlinked to the network through a full-scale converter. The generator- and grid-side converters are controlled by using an optimum proportional-integral (PI) controller. Each vector control scheme for the generator/grid-side converters has four CSA-GWO-based PI controller. In this study, the integral squared error criterion is used as an objective function. The CSA-GWO-based PI controller efficacy is validated by comparing its results with that obtained by using the genetic algorithm (GA)-based PI controller. The performance of the proposed control scheme is checked during various fault conditions. The control scheme quality is legalized by the simulation results that are achieved using MATLAB/Simulink program. With the CSA-GWO based PI controller, the dynamic and transient stability of grid-connected wind generator systems can be enhanced.

Keywords Renewable energy · Frequency converter · Variable-speed wind turbine · Hybrid cuckoo search algorithm and grey wolf optimiser

M. N. Amin
Electrical Power & Machines Department, Faculty of Engineering, Ain Shams University, Cairo, Egypt

M. A. Soliman
Dynamic Positioning and Navigation Department, Petroleum Marine Services Co., Alexandria 22776, Egypt

H. M. Hasanien · A. Y. Abdelaziz (✉)
Faculty of Engineering and Technology, Future University in Egypt, Cairo, Egypt
e-mail: ayabdelaziz63@gmail.com

1 Introduction

Increase in oil prices, depletion of fossil fuel, environmental needs and the growth in energy demand are the main reasons to transfer from traditional energy sources to renewable energy. Among several renewable energy sources, which include solar, geothermal, wind and biogas, wind energy has become the fastest growing. According to the latest statistics, it is expected that new installations of more than 55 GW will be added in the grid each year until 2023 [1].

Based on the enormous level of wind energy participation in the existing electrical grid, some problems have arisen. Some of these problems are the output power fluctuations and the low voltage through (LVRT) [2] capability. The major issue in the wind energy conversion system is the variable nature of the wind speed. The continuous changing of the wind speed led to a big impact on the output power for the wind farm. The minor change in wind speed causes significant variation in the output power. So, it is important to enhance the dynamic response of the wind farm by mitigating the fluctuations of the wind generator output power. The output power smoothing can be achieved by enhancing the control scheme or using energy storage system.

The grid disturbances such as voltage dips cause outing of wind farms to protect the apparatus from damage due to a high current flowing. The islanding of large-scale wind farms will impact on the network stability, so the wind farm should keep on connecting to the network during the low voltage for a specific time (LVRT) to support the network stability restoring. The LVRT capability can be achieved by enhancing the control scheme for the generator/grid-side converters or using energy storage system.

The system of a grid-connected wind turbine includes many components that achieve their specific functions in the energy conversion process. Wind energy conversion system (WECS) is composed of aerodynamic and mechanical parts (aerodynamic rotor and gearbox), the electrical parts (electrical generator and power electronic converter), coupling parts are the transmission lines and the transformer, and collection point, which is the point of common coupling (PCC) to the power network. The turbine is one of the most critical part used in WECS, which is employed to convert the kinetic energy of wind into mechanical energy, the generated energy depends on the wind speed. There are various types of wind turbine generator that have dominated the power system applications. These types are classified according to the orientation of the spin axis: horizontal-axis, vertical-axis [3–5]. The main difference between these two types is how the efficiency of the rotor would be limited for different wind speed conditions. The horizontal axis has many merits than the vertical axis such as blades are to the side of the turbine which is the center of gravity, helping stability, it can produce power at low cut-in wind speeds and it has the capability of self-starting due to its elevation from the tower. The electrical subsystem of the WECS consists of the generator, power converter double circuit transmission line and transformer. There are various types of generators studied with the wind turbine, among them the Squirrel-Cage Induction Generator (SCIG) [6–10],

Doubly-Fed Induction Generator (DFIG) [11–20], Wound Rotor Induction Generator (WRIG) [21–23], Permanent-Synchronous generator (PMSG) [24–34], Wound Rotor Synchronous Generator (WRSG) [35]. Those various generators are used with WECS according to the rotational speed of the rotor fixed or variable-speed wind turbine (FSWT or VSWT). Depending on the generator type, the wind turbine shaft connected to a generator through a gearbox or direct drive. So, there are various configurations of grid-connected wind generator. (i) Fixed-speed using SCIG and a three-stage gearbox [6–9], the generator connects to the wind turbine rotor through a three-stage gearbox, the fixed speed is designed to extract maximum power at certain wind speed, there are some drawbacks for this method: this type needs good mechanical design to endure the mechanical stresses, and there are high fluctuations in the output electrical power due to the wind speed variations. (ii) Limited variable-speed using WRIG and a three-stage gearbox [21–23], in this configuration the generator system connects to the wind turbine shaft through the three-stage gearbox. The power converter consists of a diode rectifier and a chopper. With adjusting the rotor speed, the torque-speed characteristic of the generator is affected. The most suitable characteristic is chosen to get the optimal operating point. The capacitor bank is used for reactive power compensation. (iii) Variable-speed [24–34] concept has many merits than FSWT as follows: reducing the noise, ability to capture the maximum power, it can reduce the mechanical stresses of the drive train of the wind turbine. The VSWT used the maximum power tracking technique (MPPT) to get the maximum output at each wind speed. The generator is coupled with the grid through power controller to adjust output frequency with the grid frequency.

There are various types of electrical generator used in the VSWG, such as (i) DFIG with a three-stage gearbox, the generator delivers power to the grid from the stator. The stator is coupled directly to the grid and the rotor connects to the grid through the power converter, and it can absorb or deliver the power to the grid according to the rotor speed, the power converter is a back-to-back converter with an intermediate DC-link capacitor (C_{DC}). The power converter consists of rotor/grid-side converters. There some drawbacks to this configuration: it has a high sensitivity to the grid disturbances, need to the gearbox, it needs frequent maintenance due to the brushes and the slip-rings, need to synchronization circuit between the stator and the grid to decrease the start-up current. (ii) PMSG has many advantages, for example, not need of power supply for its excitation circuit, it can operate at low speed, so there is not need for the gearbox, can control the active and reactive power, not need forslip-rings or brushes, so less maintenance will be required, it allows the variable-speed operation for the wind turbine so the MPPT can be achieved. The PMSG is excited by permanent magnet. The power is extracted from the stator. The generator is coupled to the grid through full-scale frequency converter (FC). The frequency converter has a critical goal such as, to work as an energy buffer, enable to control active and reactive power, adjust the terminal and DC-link voltage to be constant at different operation conditions.

In this study the model system is composed of grid-tied VSWT direct drive PMSG, and the generator is connected to the grid through full-scale FC. The FC composes of a force-commutated rectifier and a force-commutated inverter, and each is composed of six IGBTs connected with common C_{DC}. The ac voltage of the PMSG is converted to dc which is connected to the DC-link at the generator side converter (GSC) and then converted to ac at the grid-side inverter (GSI) with a certain frequency according to the network frequency. So, the GSC works as a rectifier and the GSI works as an inverter. Each control scheme for GSC and GSI consists of four optimum proportional-integral controller. The GSC is capable to extract maximum power from the generator to the grid at unity power factor and the GSI has the responsibility to adjust the terminal voltage (V_{PCC}) and the DC-link voltage (V_{DC}) at certain value during the different operation conditions. The C_{DC} is protected by using over voltage protection scheme (OVPS) from damage during the fault. There is another coupling component to the grid such as two parallel transmission line and set-up transformer.

In this study, the vector control depends on the PI [30–34, 36–40] controller. The proportional-derivative (PD) and proportional-integral-derivative (PID) controllers are not more attractive to be used in the industry due to the demerits of derivative control action, where it amplifies the input frequency of any harmonics to the system, and to avoid this problem, a designed filter should be used. Therefore, PI controller is commonly used in the industry. PI controllers have characterized by having a large stability limit, although they are affected by the variation of the parameters. Four PIs are utilized for each converter, so the fine-tuning of these controllers is arduous, especially in the nonlinear system.

Classical PI controller is used in many applications due to their robust, strength and stabilization. Generally, a cascaded control scheme is considered as a main controller technique that can be used in the converter systems. It consists of two loops, inner and outer loop, where each loop consists of two PI controllers. This control structure is very complicated and consumes a long time.

Many approaches are utilized to reach the optimal parameters of the PI controller. Traditionally, Newton-Raphson (NR) and Ziegler-Nichols (ZN) approaches are used to fine-tune the PI controller, but these optimizations methods are affected by the type of solver and the initial conditions. Besides, the heuristic approaches, such as the gravitational search algorithm (GSA) [41], a genetic algorithm (GA) [42], the particle swarm algorithm (PSO) [43], a grey wolf optimizer (GWO) [44], and the ant colony optimization (ACO) are used to obtain the optimal values of the PI controllers. But these methods suffer from some drawbacks in finding the ideal solution. The GA technique can be applied for both continuous and discrete parameters, but it requires a complex analysis, a long time, and can be trapped into local optima. The PSO has some advantages such as its simplicity and it has few parameters to adjust. But, PSO can have difficulty to get the initial design parameters, and it can converge early and can be trapped into local optimum solutions with complex problems. GSA depends on Newton's law of mass attraction and gravity. GSA has some merits, including the ability to solve highly nonlinear optimization problems and has stable convergence characteristics. But, GSA has drawbacks such as slow searching speed in the last iterations and long computational time. ACO can give rapid discovery of reasonable solutions and has guaranteed convergence. But, ACO has an uncertain time to convergence and has a complicated theoretical analysis. The GWO has some demerits, such as slow convergence and inadequate local searching ability.

The main characteristics of the optimization process are the balance between the exploration and the exploitation to reach the optimal solution. This article presents the application of cuckoo search algorithm and grey wolf optimizer (CSA-GWO) [45]-based PI to control the FC. The GWO is inspired by the unique predator strategy and organization system of grey wolves. Since the GWO algorithm is easy to fall into local optimum especially when it deals with high dimensional data. By using global search ability of CSA [46, 47] in GWO to update the solution of GWO, the search ability of GWO is strengthened, and the shortcoming of GWO is offset. The CSA-GWO has many merits, such as better capability to escape from local optima solution and has a faster convergence rate. The performance of CSA-GWO is ensured by comparing the results that are achieved by CSA-GWO-based PI controller with that obtained from the application of a GA-based PI controller. The various types of faults are considered in this comparison. These simulation results are performed by utilizing the MATLAB/Simulink programming. With the CSA-GWO-based PI controller, the dynamic and transient stability of grid-connected wind generator systems can be enhanced.

2 Model System

Figure 1 shows the model system of the grid-tied VSWT direct drive PMSG. The output power of the PMSG is controlled by using full-scale FC. The components of the FC are two full-scale power converters at the generator and grid-side, C_{DC}. The capacitor is protected by an OVPS during the fault. The OVPS consists of an insulated gate bipolar transistor (IGBT) connected in series with a resistance. The ac output voltage of the PMSG is converted to dc voltage at the GSC and linked with C_{DC} and then converted to ac voltage with adjusted frequency according to network frequency at the GSI. So, the grid-side converter works as a rectifier and the grid-side inverter work as an inverter. Each vector control scheme has four PI controllers to control the generator and grid-side converters. The GSC is managed to extract maximum power at unity power factor from the generator. The V_{PCC} and the DC voltage V_{DC} are regulated by the GSI. The VSWT-PMSG is coupled with the network through the step-up transformer and double circuit transmission line. The parameters of PMSG are demonstrated in Table 1. The system base is considered as 1.5MVA.

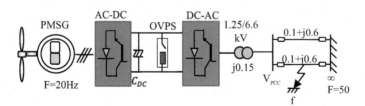

Fig. 1 System model

Table 1 Generator parameters

P_{rated}	1.5 MW	R_S	0.006 pu
V_{rated}	275 V	X_d	0.7 pu
Frequency	20 Hz	X_q	0.3 pu
Poles No.	48	Field flux	0.9 pu

3 Wind Turbine Model

The relation among the wind speed and mechanical power can be presented as follows [24]:

$$P_\omega = 0.5 \, \rho\pi \, R^2 V_\omega^3 C_P(\lambda, \beta) \tag{1}$$

where V_ω is the speed of the wind, ρ is the density of air, R is the blade radius, C_P presents the power coefficient, which depends on turbine pitch angle, β and tip speed ratio, λ and it can be modeled as:

$$\lambda = \frac{\omega_r R}{V_\omega} \tag{2}$$

$$L_i = \frac{1}{\left(\frac{1}{\lambda + 0.02\beta}\right) - \left(\frac{0.03}{\beta^3 + 1}\right)} \tag{3}$$

$$C_p(\lambda, \beta) = 0.73\left(\left(\frac{151}{Li}\right) - 0.58\beta - 0.002\left(\beta^{2.14}\right) - 13.2\right)e^{-18.4/L_i} \tag{4}$$

where ω_r means the rotational rotor speed.

VSWT is preferred over the FSWT in WECS due to its ability for maximizing the extracted wind power under wind speed variations. The value of the power coefficient can be controlled by the value of tip speed ratio, which maximizes the capture power at any speed. This technique is defined as maximum power point tracking (MPPT). Figure 2 exhibits the dynamic characteristic of the turbine. It is noted that there are different curves for the extracted power at various wind speeds. For each curve, a specific point where the maximum power can be captured. So, the adjustable controller has to follow the particular rotational speed to get the maximum power at each wind speed.

To get P_{max} in terms of ω_r utilize the following equation [3]:

$$P_{max} = 0.5\rho\pi R^2 \left(\frac{\omega_r R}{\lambda_{opt}}\right)^3 C_{P-opt} \tag{5}$$

Fig. 2 Wind turbine characteristic with MPPT curve

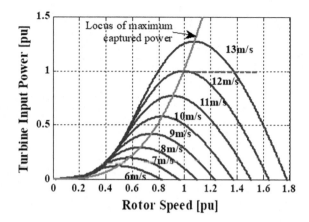

where C_{p-opt} is the coefficient of optimum power and λ_{opt} is the optimum value of the tip speed ratio. The rated power will be otherwise called the optimum power, P_{opt} which is utilized as the controlling point for the power at FC.

4 Control of FC

The FC is utilized in controlling the behavior of the grid-connected VSWT-PMSG, to extract P_{max} from the generator and adjusting the V_{PCC}, V_{DC} at various operating conditions. It composes of the GSI, GSC, and intermediate capacitor.

4.1 Generator-Side Converter (GSC)

The GSC has the responsibility to deliver P_{max} from the generator at power factor equal one. This output is obtained by adjusting the dq-axes components of the stator current (I_d, I_q).

The P_{max} from the generator is achieved by adjusting the I_q according to the set-point (P_{opt}), which is obtained from MPPT. The I_d is adapted to deliver the power at unity power factor. As illustrated in Fig. 3, the controlling of the GSC is based on four PI controllers to create the reference voltage in the dq-axes components $(V_{d,q}^*)$. These components are transformed into three-phase components $(V_{a,b,c}^*)$. A comparison is occurred between $V_{a,b,c}^*$ and a triangular carrier waveform (TCW) with 1.0 kHz, to get firing pulses (FP) for the converter. The transformation from dq-axes to abc-axes frame is occurred by utilizing the transformation angle (θ_r), which is elicited from the ω_r.

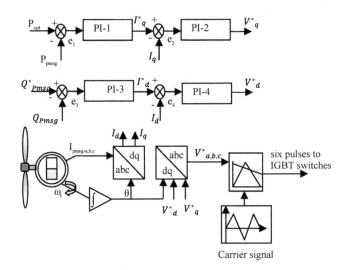

Fig. 3 The GSC block diagram

4.2 Grid-Side Inverter (GSI)

The GSI has the responsibility to control V_{DC} and V_{PCC} by adjusting the d-axis and q-axis grid current components (I_{dn}, I_{qn}). The I_{qn} is adjusted to control the V_{DC}, and the I_{dn} is regulated to maintain the V_{PCC} to be constant at severe disturbances. In this analysis, the C is selected as 10 mF with a rated voltage with rate value equal to 1.15 kV. As illustrated in Fig. 4, the grid-side inverter control block diagram, which consists of four PI controllers, are utilized to extract the reference voltage ($V_{dn,qn}^*$). The $V_{dn,qn}^*$ are transformed into three-phase signals $V_{an,bn,cn}^*$ by utilizing a transformation angle (θ_t). A comparison is occurred between $V_{an,bn,cn}^*$ with a TCW with 1050 Hz to produce FP to the inverter. To extract (θ_t), a phase-locked-loop (PLL) is utilized.

4.3 Over Voltage Protection Scheme (OVPS)

During the fault condition, the reactive power (Q_r) is delivered to the grid. At the same time, no real power (P_r) is transmitted to the network, which results in increasing the V_{DC} and leading to destroy the C_{DC}. The OVPS is needed to protect the C from damaging. The components of the OVPS are the controlled IGBT with series in resistance [30, 31]. When V_{DC} exceeds 20% above the nominal value of the V_{DC}, IGBT will operate to dissipate the power through the resistance. So, OVPS will avoid the destruction of the intermediate C_{DC}.

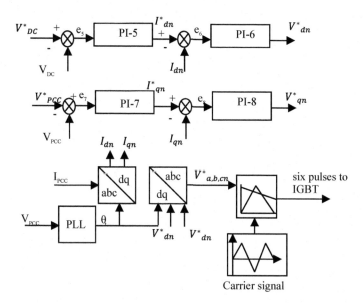

Fig. 4 The GSI block diagram

5 Optimization Design of Cascaded PI Controller

5.1 GWO Algorithm

GWO is considered one of the recent algorithms in the meta-heuristic optimization technique and is highly useful to deal with the optimization problems. GWO algorithm mimics the lifestyle of the grey wolves in nature in connection with hierarch of society and hunting techniques. Due to the clear division of labor, the population of grey wolf is divided into four hierarchical systems. The selected types of grey wolves are α, β, δ and ω. The main steps of the hunting process for optimization execution are searching and tracking the prey then encircling and harassing the prey, and eventually attack the prey. There are many merits in GWO that are more flexible, simple. The first step in the GWO optimization procedure is the creation of a random population of grey wolves. In each iteration the possible position of the prey can be estimated by α, β, δ wolves to represent the optimal solution. The position of the grey wolves is updated according to the distances between the grey wolves and the prey.

5.2 CSA

CSA is one of the meta-heuristic algorithms, which is inspired by nature. CSA is a stochastic optimization algorithm, which is based on population. CSA is fast and highly capable algorithm. The CSA can look for the best optimum value through the specified search space in a better approach than other techniques. As the CSA has memory mechanization for helping to record the local minimal and participate to pick out the best. The CSA mimics the lifestyle and reproductivity for a sort of bird called cuckoo bird. The basic steps of CSA are as follows:

1- Each cuckoo puts one egg at a time in a randomly selected nest.
2- The best nests of high-quality eggs only will be transferred to the following generation.
3- The host bird may discover the foreign egg of a cuckoo.

5.3 CSA-GWO

The hybridization of two algorithms is recently emerging to find superior solutions to the optimization problems. The main characteristics of the optimization process are the balance between the exploration and the exploitation to reach the optimal solution.

GWO inspects an individual with a high fitness value, and accordingly has a weak ability to get the global optimum solution, and easy to fall-into the local optimum. By using global search ability of CSA in GWO to update the solution of GWO, the search ability of GWO is strengthened, and the shortcoming of GWO is offset. According to this, CSA is a very helpful tool for GWO improvement. The flowchart for CSA-GWO is described in Fig. 5.

The key group parameters in GWO are updated by cuckoo search's position updating formula as shown in the flow chart. The position updated equation of the CSA is applied to update the position, speeds and convergence behaviors of the grey wolf agent (α) to balance between the exploring, exploiting and expanding convergence behaviors of the GWO algorithm. The remaining process of the GWO algorithm is considered as it is. The analytical solution obtained using the new hybrid CSA-GWO-based PI controller is compared with another meta-heuristics approach GA-based PI controller to examine the efficiencies of the design parameters.

5.4 Problem Formulation

Choosing of the objective function in our problem is done by using integral square error (ISE) criterion, as expressed:

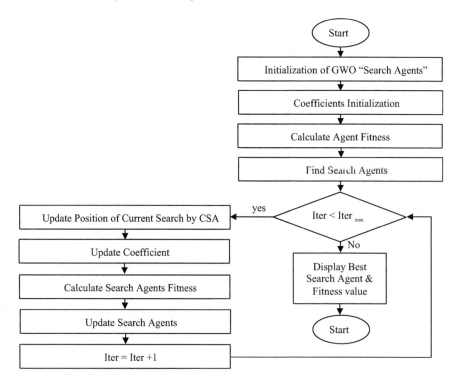

Fig. 5 Flow chart of CSA-GWO algorithm

$$ISE = \int_0^t ((e_1^2 + e_2^2 + e_3^2 + e_4^2) + (e_5^2 + e_6^2 + e_7^2 + e_8^2)).dt \qquad (8)$$

where e_1, e_2, e_3 and e_4 are the actuating errors of P_{opt}, I_q, Q_{pmsg} and I_d, respectively, for generator-side and e_5, e_6, e_7 and e_8 are the actuating errors of V_{DC}, I_{dn}, V_{PCC} and I_{qn}, respectively, for the grid-side.

6 Simulation Results

The simulation results are achieved by utilizing the MATLAB/Simulink program. The time step in the simulation is five micro sec. The results of the CSA-GWO-based PI controller are compared with that are produced by the GA-PI controller to confirm the CSA-GWO-based PI efficacy under the various disturbances.

The three-line to ground fault (3LG) is occurred at 1.7 s at point f, as shown in Fig. 1. At 1.8 s, the circuit breaker at the faulted line is opened and then reclosed at 2.1 s after clearing the fault.

Figure 6 shows the response of the active/reactive power, which is transmitted to the network, and the terminal/DC-link voltage by using the CSA-GWO-based PI controller and GA-based PI controller, when the 3LG is considered. Figure 6a illustrates the response of the P_r, which is transmitted to the network. It is notable, the CSA-GWO-based PI controller helps the P_r response to reach rapidly to its pre-fault value. As shown in Fig. 6b, the response of the Q_r is improved by using CSA-GWO-based PI controller than that is obtained by the GA-PI. Notably, low

Fig. 6 Responses for 3LG temporary fault **a** P_r. **b** Q_r. **c** V_{DC}. **d** V_{PCC}

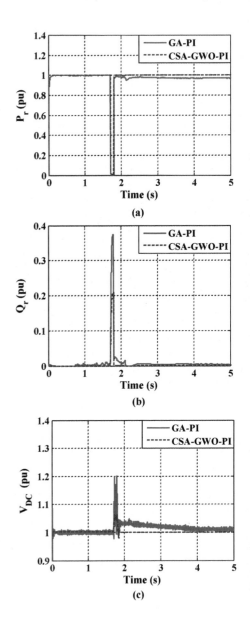

Fig. 6 (continued)

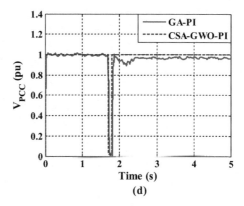

(d)

fluctuations and low over-shoot in the response of the Q_r is obtained by using CSA-GWO-based PI controller. Figure 6c illustrates the response of V_{DC}. It is noticed that the response of V_{DC} has a better-damped response by utilizing CSA-GWO-based PI controller than that is achieved by GA-based PI controller. Figure 6d presents the V_{PCC} response. Notably, at the steady-state, the response of V_{PCC} using CSA-GWO-based PI controller has low settling time and has a small steady state error than that is obtained by utilizing GA-based PI controller.

For more verification of the proposed methodology, unsymmetrical fault conditions are considered in the analysis, such as single-line-to-ground (LG), line-to-line-to-ground (LLG) and line-to-line (LL), utilizing both control approaches. Figure 7a–c describe the response of the P_r, Q_r, V_{PCC} under (LLG). Figure 8a–c describe the response of the P_r, Q_r, V_{PCC} under (LL). Figure 9a–c describe the response of the P_r, Q_r, V_{PCC} under (LG).

The simulation results prove the efficiency of using CSA-GWO-based PI controller in enhancing the transient characteristics of the grid-tied VSWT-direct drive PMSG. The CSA-GWO-based PI can obtain accurate, reliable and satisfactory responses compared with the using GA-based PI controller in the wind energy conversion system.

7   Conclusion

This article has exhibited an implementation of the CSA-GWO-based PI controller to improve the behavior of the grid-tied VSWT-direct drive-PMSG. The CSA-GWO-based PI is legalized by comparing its results with that achieved by utilizing GA-based PI controller. The suggested strategy is utilized in controlling the FC to extract maximum power from the PMSG at unity power factor, then transmitted P_r with enhanced transient characteristics to the network, and to adjust the terminal/DC-link voltage to be constant at various operating conditions. The comparison between simulation results of the CSA-GWO-based PI controller and the GA-based PI controller

Fig. 7 Responses for LLG
temporary fault **a** P_r. **b** Q_r.
c V_{PCC}

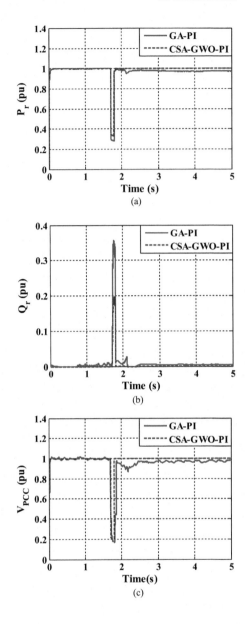

is used to confirm the effectiveness of the proposed control strategy, taking into consideration subjecting the system to various network disturbances. It is clear from the simulation results that the transient system responses using CSA-GWO-based PI controller have better-damped responses and minimum oscillations than that are obtained using GA-based PI. The efficiency of the CSA-GWO-based PI is checked under various fault conditions. The CSA-GWO-based PI controller has the efficiency

Fig. 8 Responses for LL temporary fault **a** P_r. **b** Q_r. **c** V_{PCC}

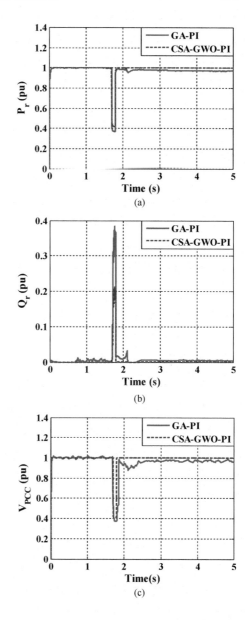

(a)

(b)

(c)

to deal with controlling the nonlinear system. Therefore, the proposed control scheme can be used to obtain enhanced response in other renewable energy systems, smart grids and storage energy systems.

Fig. 9 Responses for LG
temporary fault **a** P_r. **b** Q_r.
c V_{PCC}

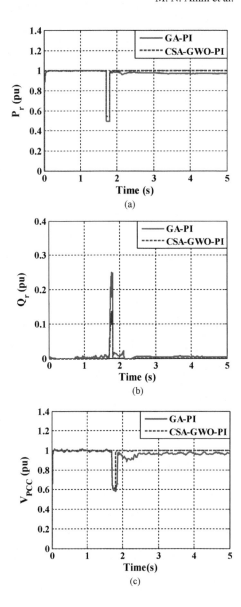

References

1. Global Wind Energy Council (GWEC), Global Wind Energy Outlook (2017). http://www.gwe
   c.net
2. Qais MH, Hasanien HM, Alghuwainem S (2017) Low voltage ride-through capability enhancement of grid-connected permanent magnet synchronous generator driven directly by variable
   speed wind turbine: a review. J Eng 2017(13):1750–1754

3. Maura S, Hara N, Akimoto H, Konishi K (2019) Torque control of the floating vertical axis wind turbine with a frictional power take-off system. In: 2019 international automatic control conference (CACS), Keelung, Taiwan, pp 1–6
4. Gumilar L, Kusumawardana A, Prihanto D, Wicaksono H (2019) Analysis performance vertical axis wind turbine based on pitch angle to output power. In: 2019 international conference on information and communications technology (ICOIACT), Yogyakarta, Indonesia, pp 767–772
5. Kusumawardana A, Gumilar L, Prihanto D, Wicaksono H, Saputra SNT, Prasetyo D (2019) Simple MPPT based on maximum power with double integral sliding mode current control for vertical axis wind turbine. In: 2019 IEEE conference on energy conversion (CENCON), Yogyakarta, Indonesia, pp 31–36
6. Firouzi M, Gharehpetian GB (2017) A modified transformer-type fault current limiter for enhancement fault ride-through capability of fixed speed-based wind power plants. In: 2017 conference on electrical power distribution networks conference (EPDC), Semnan, pp 33–38
7. Dhouib B, Kahouli A, Abdallah HH (2017) Dynamic behavior of grid-connected fixed speed wind turbine based on proportional-integral pitch controller and fault analysis. In: 2017 international conference on green energy conversion systems (GECS), Hammamet, pp 1–7
8. Hazari MR, Umemura A, Takahashi R, Tamura J, Mannan MA (2017) A new fuzzy logic based control strategy for variable speed wind generator to enhance the transient stability of fixed speed wind generator. In: 2017 IEEE manchester powertech, Manchester, pp 1–6
9. Lee C, Chen L, Tsai S, Liu W, Wu Y (2009) The Impact of SCIG wind farm connecting into a distribution system. In: 2009 Asia-Pacific power and energy engineering conference, Wuhan, pp 1–7
10. Hasanien HM, Muyeen SM (2012) Speed control of grid-connected switched reluctance generator driven by variable speed wind turbine using adaptive neural network controller. Electr Power Syst 84(1):206–213
11. Hasanien HM, Al-Ammar EA (2012) Dynamic response improvement of doubly fed induction generator-based wind farm using fuzzy logic controller. J Electr Eng 63(5):281–288
12. Errouissi R, Al-Durra A, Muyeen SM, Leng S, Blaabjerg F (2017) Offset-free direct power control of dfig under continuous-time model predictive control. IEEE Trans Power Electron 32(3):2265–2277
13. Zhu X, Pan Z (2017) Study on the influencing factors and mechanism of SSR due to DFIG-based wind turbines to a series compensated transmission system. In: 2017 IEEE 26th international symposium on industrial electronics (ISIE), Edinburgh, pp 1029–1034
14. Hiremath R, Moger T, Comparison of LVRT enhancement for DFIG-based wind turbine generator with rotor-side control strategy. In: 2020 international conference on electrical and electronics engineering (ICE3), Gorakhpur, India, pp 216–220
15. Zhu L, Hu X, Li S (2018) High-frequency resonance of DFIG-based wind generation under weak power network. In: 2018 international conference on power system technology (POWERCON), Guangzhou, pp 2719–2724
16. Jiao Y, Nian H, He G (2017) Control strategy based on virtual synchronous generator of DFIG-based wind turbine under unbalanced grid voltage. In: 2017 20th international conference on electrical machines and systems (ICEMS), Sydney, NSW, pp 1–6
17. Chabani MS, Benchouia MT, Golea A, Boumaaraf R (2017) Implementation of direct stator voltage control of stand-alone DFIG-based wind energy conversion system. In: 2017 5th international conference on electrical engineering—Boumerdes (ICEE-B), Boumerdes, pp 1–6
18. Vishwanath GM, Padhy NP (2018) Power hardware in loop based experimentation on DFIG wind system integrated to isolated DC microgrid. In: 2018 IEEE international conference on power electronics, drives and energy systems (PEDES), Chennai, India, pp 1–5
19. Liu B, Hu Y, Su W, Li G, Wang H, Li Z (2019) Stability analysis of DFIG wind turbine connected to weak grid based on impedance modeling. In: 2019 IEEE power & energy society general meeting (PESGM), Atlanta, GA, USA, 2019, pp 1–5
20. Peng T, Ze-Tao L, Zheng-Hang H, Shi Z, Jun-Xian H (2018) Experimental research on grid-connected control system of DFIG with uncertain parameter. In: 2018 Chinese Automation Congress (CAC), Xi'an, China, pp 3350–3354

21. Sowmmiya U, Govindarajan U (2018) Control and power transfer operation of WRIG-based WECS in a hybrid AC/DC microgrid. In: IET renewable power generation, vol 12, no 3, pp 359–373
22. Najafi HR, Robinson FVP, Dastyar F, Samadi AA (2009) Small-disturbance voltage stability of distribution systems with wind turbine implemented with WRIG. In: 2009 international conference on power engineering, energy and electrical drives, Lisbon, pp 191–195
23. Zhang S, Tseng KT, Nguyen TD (2008) WRIG based wind conversion system excited by matrix converter with current control strategy. In: 2008 IEEE International Conference on Sustainable Energy Technologies, Singapore, pp 203–208
24. Chen J, Lin T, Wen C, Song Y (2016) Design of a unified power controller for variable-speed fixed-pitch wind energy conversion system. IEEE Trans Industr Electron 63(8):4899–4908
25. Zhang Z, Zhao Y, Qiao W, Qu L (2014) A space-vector-modulated sensorless direct-torque control for direct-drive PMSG wind turbines. In: IEEE transactions on industry applications, vol 50, no 4, pp 2331–2341
26. Choi JW, Heo SY, Kim MK (2016) Hybrid operation strategy of wind energy storage system for power grid frequency regulation. In: IET generation, transmission & distribution, vol 10, no 3, pp 736–749
27. Kassem AM (2016) Modelling and robust control design of a standalone wind-based energy storage generation unit powering an induction motor-variable-displacement pressure-compensated pump. In: IET renewable power generation, vol 10, no 3, pp 275–286
28. Qais MH, Hasanien HM, Alghuwainem S (2018) A grey wolf optimizer for optimum parameters of multiple PI controllers of a grid-connected pmsg driven by variable speed wind turbine. IEEE Access 6:44120–44128
29. Soliman MA, Hasanien HM, Azazi HZ, El-Kholy EE, Mahmoud SA (2019) An adaptive fuzzy logic control strategy for performance enhancement of a grid-connected PMSG-based wind turbine. IEEE Trans Ind Infor 15(6):3163–3173
30. Amin MN, Soliman MA, Hasanien HM, Abdelaziz A (2020) Grasshopper optimization algorithm-based pi controller scheme for performance enhancement of a grid-connected wind generator. J Control Autom Electr Syst 31:393–401
31. Amin MN, Soliman MA, Hasanien HM, Abdelaziz A, Ali Z (2019) Grasshopper optimization algorithm-based fuzzy logic controller for performance enhancement of a grid-connected wind generator. J Energy Conver (*IRECON*), 7(6):218–229
32. Soliman MA, Hasanien HM, Al-Durra A (2019) High-performance frequency converter controlled variable-speed wind generator using linear-quadratic regulator controller. In: 2019 IEEE industry applications society annual meeting, Baltimore, MD, USA, pp 1–7
33. Soliman MA, Hasanien HM, Al-Durra A, Alsaidan I (2020) A novel adaptive control method for performance enhancement of grid-connected variable-speed wind generators. IEEE Access 8:82617–82629
34. Soliman MA, Hasanien HM, Alghuwainem S, Al-Durra A (2019) Symbiotic organisms search algorithm-based optimal control strategy for efficient operation of variable-speed wind generators. In: IET renewable power generation, vol 13, no 14, pp 2684–2692, 28
35. Taher SA, Karimi MH, Arani ZD (2019) Improving fault ride through capability of full-Scale WRSG wind turbines using MPC-based DVR. In: 2019 27th Iranian conference on electrical engineering (ICEE), Yazd, Iran, pp 863–867
36. Hasanien HM, Muyeen SM (2012) Design optimization of controller parameters used in variable speed wind energy conversion system by genetic algorithms. IEEE Trans Sustain Energy 3(2):200–208
37. Qais MH, Hasanien HM, Alghuwainem S (2017) Output power smoothing of grid-connected permanent-magnet synchronous generator driven directly by variable speed wind turbine: a review. J Eng 2017(13):1755–1759
38. Qais MH, Hasanien HM, Alghuwainem S (2019) Enhance salp swarm algorithm: application to variable speed wind generators. Eng Appl Artif Intell 80:82–96
39. Nityanand, Pandey AK (2018) Performance analysis of PMSG wind turbine at variable wind speed. In: 2018 5th IEEE Uttar Pradesh section international conference on electrical, electronics and computer engineering (UPCON), Gorakhpur, pp 1–6

40. Gupta RA, Singh B, B Jain (2015) Wind energy conversion system using PMSG. In: 2015 international conference on recent developments in control, automation and power engineering (RDCAPE), Noida, pp 199–203
41. Sharma A, Mathur S, Deterministic maximum likelihood direction of arrival estimation using GSA. In: 2016 international conference on electrical, electronics, and optimization techniques (ICEEOT), Chennai, pp 415–419
42. Odofin S, Gao Z, Sun K (2015) Robust fault estimation in wind turbine systems using GA optimisation. In: 2015 IEEE 13th international conference on industrial informatics (INDIN), Cambridge, pp 580–585
43. Hasanien HM, Muyeen SM (2015) Particle swarm optimization-based superconducting magnetic energy storage for low-voltage ride-throughcapability enhancement in wind energy conversion system. Electr Power Compon Syst 43(11):1278–1288
44. Yadav Y, Parmar G, Bhatt R (2019) Application of GWO with different performance indices for BH system. In: 2019 4th international conference on information systems and computer networks (ISCON), Mathura, India, 2019, pp 742–745
45. Mahmoud HY, Hasanien HM, Besheer AH, Abdelaziz AY (2020) Hybrid cuckoo search algorithm and grey wolf optimiser-based optimal control strategy for performance enhancement of HVDC-based offshore wind farms. In: *IET* generation, transmission & distribution, vol 14, no 10, pp 1902–1911
46. Kalaam RN, Muyeen SM, Al-Durra A, Hasanien HM, Al-Wahedi K (2017) Optimisation of controller parameters for grid-tied photovoltaic system at faulty network using artificial neural network-based cuckoo search algorithm. In: IET renewable power generation, vol 11, no 12, pp 1517–1526
47. Zefan C, Xiaodong Y (2017) Cuckoo search algorithm with deep search. In: 2017 3rd IEEE international conference on computer and communications (ICCC), Chengdu, pp 2241–2246

Wind Energy System Grid Integration and Grid Code Requirements of Wind Energy System

Kishor V. Bhadane, Tushar H. Jaware, Dipak P. Patil, and Anand Nayyar

1 Introduction

Energy is indeed an integral element for daily life. Energy has become a highly vital utility for every country and also its availability will impact the entire output of financial resources in a dynamic manner [33].

Energy is produced in numerous ways; however electrical energy is among the major sources. Modern world community is heavily dependent aon the usage of electrical energies. This makes it an important part of our lives. Therefore, electricity has a prime role in industrial civilization. The fundamental need for civilization and the industrial sector is to thrive the reasonably cheap and constant production of electrical energy. The benefit of electrical power has rendered it preferable to several other sources of electricity [20].

In India, for vast quantities of power generation, natural energy options such as fossil fuels, nuclear energy, hydropower, etc., have been utilized. Because of these resources, the climate is adversely affected; these sources will also be costly and

K. V. Bhadane (✉)
Electrical Engineering, Amrutvahini College of Engineering, Ahemednagar, Sangamner, Maharashtra, India
e-mail: kishor4293@yahoo.co.in; kishor.bhadane@avcoe.org

T. H. Jaware
Electronics and Telecommunication Engineering, R C Patel Institute of Technology, Shirpur, India
e-mail: tusharjaware@gmail.com

D. P. Patil
Electronics and Telecommunication Engineering, Sandip Institute of Engineering and Management, Nashik, India
e-mail: Dipak.patil@siem.org.in

A. Nayyar
Duy Tan University, Da Nang, Viet Nam
e-mail: anandnayyar@duytan.edu.vn

© The Author(s), under exclusive license to Springer Nature Switzerland AG 2021
A. M. Eltamaly et al. (eds.), *Control and Operation of Grid-Connected Wind Energy Systems*, Green Energy and Technology,
https://doi.org/10.1007/978-3-030-64336-2_10

stressful in the years ahead. The cost reduction of solar green energy is achieved due to technical improvement [26].

Wind and solar power, in grid-connected or isolated dispersed form, are the globally used renewable and alternative energy resources [1]. Wind energy is one of the renewable energy resources that save our environment from being polluted.

Outline of the Chapter

This chapter focuses on the actual and potential energy scenarios for the existence and usage of different forms of renewable and non-renewable energy supplies in India compared to other nations. There is a necessity of wind energy, the development of the power sector, main challenges to production for wind energy, issues regarding implementation, the importance of the quality of electricity and customized power devices were addressed.

It represents wind energy technologies, together with its grid codes, and various subcomponents and categories. The various wind power system topologies are explained. The technological specifications regarding power quality problems for the wind turbine system are being specified according to the wind grid codes.

2   Energy Development

In the energy sector, Bharat is the largest customer worldwide, after China, the US, and Russia. India's energy demand is increasing at an annual growth rate of 2.8%, owing to its continuing growth in the economy [34].

In order to reduce the disparity between demand and the availability of energy resources, substantial measures are needed to boost energy supplies.

India's budget for importing oil is almost 80%, so energy-related security problems will be emphasized in the near future. The major portion of electricity generation in India is from traditional sources of energy and raw fuels are limited. It thus impacts the production of electricity and becomes more highly dependent on imports of coal. Similarly, besides the power generation, the supply of gas is limited to address the need for fuel. Through the use of nuclear power and massive hydropower plants, there are many significant issues mostly for the production of electricity. Therefore, power cuts and load shedding are a growing concern. Due to modern civilization, economic growth, energy demand is growing rapidly; per capita consumption has increased too with income and further energy usage. It therefore adds to the imbalance between energy supplies and demand. Due to power shortages, several industries such as domestic, commercial, and industrial use more diesel as well as furnace oil as backup to satisfy their demand for power. The challenge of load shedding is confronted by the rural population who are using kerosene as an interim solution to deal with this issue. This results in a rise within the expense of different subsidies and thereby limits the economic growth of the country's oil imports.

Even after several years of Indian independence, village's citizens often experience a lack of electricity available. As electricity consumption has not risen with

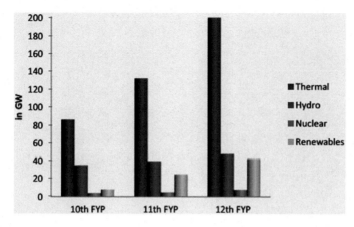

Fig. 1 Installed capacity in different five-year plans (*Source* EY Analysis) (*Source* MNRE, GoI; CEA Statistics)

population growth, there is a potential to balance the demand and power supply gap. This can be accomplished by significantly rising the potential of natural renewable energy production [35].

3   Enhancement of Power Zone

3.1   Power Development in India

India is ranked fifth in the universe in power production. India's current capacity for power generation is around 60%, especially for coal. India relies on importing oil and coal to meet the demand for electricity. In the recent years India's installed capacity has increased significantly to 272.5 GW [27]. In terms of energy consumption, India achieved fourth place in the world compared with 7th in 2000 [29].

During the five-year plan and the twelfth five-year plan, accumulated installed capacity is stated in Fig. 1. There are about 45GW of renewable electricity.

Figure 2 illustrates India's current energy scenarios. It is stated that the overall installed and renewable energy capacities are 264 and 35 GW. The renewable energy is 13.1 GW, out of which 8.6 GW is wind energy.

3.2   Progressive Status of Renewable Energy (RE) in India

By May 2015, the production of electricity in India is 272.5 GW and out of that the renewable energy power is 35.8 GW [28]. The Indian renewable energy [RE] capacity

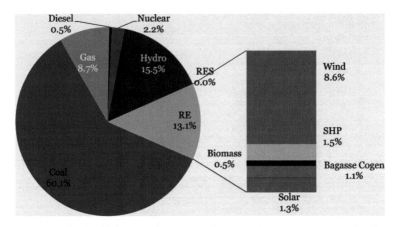

Fig. 2 Existing power development of India in 2015

is currently valued at around 895 GW [29]. This ability needed to be developed then utilized. This potential lies in solar, wind, biomass, and hydro types that are abundant in our nation in massive quantities. There is tremendous wind and solar capacity in Maharashtra, Gujarat, Tamilnadu, Andhra Pradesh, and Karnataka, to name only a few. Figure 3 reveals the RE Installed Capacity in India (May 2015). Until May 2015, the contributions of wind power were reported as 64%.

India's global position in RE installed capacity
Figure 4 illustrates the status of the globally deployed RE power.

The World Wind Power potential in GW can be seen in Fig. 5. It shows that, with a total installed wind energy capacity of 114.6 GW, China is playing a leading role. With an installed wind energy capacity of 22.6 GW, India is in the fifth position.

RE capability in India.

Fig. 3 RE existing capability in India (May 2015) [1]

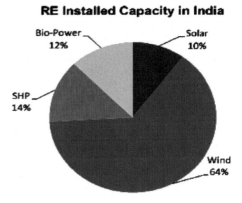

Fig. 4 Global RE installations

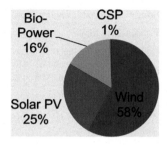

Fig. 5 Country wise wind power in GW (*Source of Figs.* 4 and 5: As on Dec 2014: Global Wind Energy Council, As on Jan 2014, IEA PVPS)

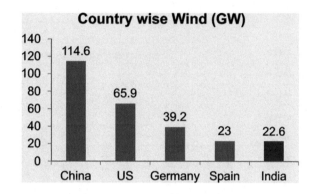

The source wise renewable capacity of 9389–34351 MW from 2007 to 2015 is depicted in Fig. 6. The wind energy production in 2007 was 5200 MW. While it reached 22645 MW by the end of March 2015. The renewable energy capacity of MW is represented in Table 1 including its status and updated targets. Renewable energy is 24,914 MW at the ending of the 11th plan (March 2012), while the present deployed renewable capacity is 34351 MW prior to March 2015. In the 12th plan (March 2017), the renewable energy target is 54914 MW. The revised RE target was expected to be 1,75,000 MW by 2022.

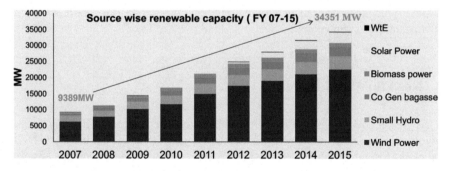

Fig. 6 RE capacity according to source

Table 1 Renewable energy capacities in MW

Capacities in MW

Source	Installed capacity by end of 11th plan (March 2012)	Current installed capacity (March 2015)	Target as per 12th plan (March 2017)	Revised targets till 2022
Solar power	941	3383	10,941	100,000
Wind power	17,352	22,654	32,352	60,000
Biomass power	3225	4183	6125	10,000
Small hydro	3395	4025	5495	5000
Total	24,914	34,351	54,914	175,000

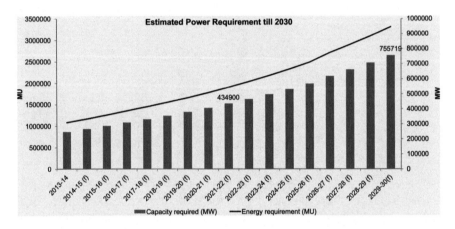

Fig. 7 Predictable power capability and requirements (*Source* Central Electricity Authority)

Predictable power obligation
The energy demand is predicted in Fig. 7.

Wind energy across states: As specified by the Center for Wind Energy Technology (CWET) guidelines. The power of Indian wind energy is 22645 MW. The distribution of wind energy is shown in Fig. 8.

4   Need for Wind Power

It has greater advantages for the environment. There is no pollution. It doesn't need any gasoline. In remote areas, distributed power generation is possible [2, 11]. At the global level the installed RE is 674 GW and out of which wind energy capacity is 371 GW. India, with a capacity of 23 GW at the end of November 2015, is in the fifth place [36].

Fig. 8 Wind energy across states (*Source* Ministry of New and Renewable Energy)

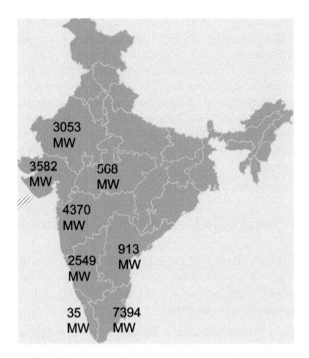

Key obstacles to Wind Energy Development:

- A high investment cost is a major challenge in the development of wind energy.
- Scarcity in the supply of financial lending and high interest rates are significant issues for the wind industry.
- In rural areas, the incorporation of wind energy into the poor grid poses technological difficulties such as power quality problems. In lower wind velocity areas, wind energy system output may be impaired.
- The wind sector faces infrastructure and supply chain problems as well.
- Due to various low import taxes on component parts of wind energy relative to raw materials utilized in the wind industry, people prefer to manufacture component parts of wind energy over local production components.

5 Significance of Electrical Power Quality in Wind Power System

Power quality is defined as power system load capabilities work satisfactorily under normal operating conditions without affecting its power output [2, 7, 8, 19]. The deployment of renewable power into the grid is accountable for impacting the wind energy system's power output [10]. The evaluation of power quality problems in such disruptions is thus necessary as well as the effect of low power quality on the health and economy of different power system equipment and networks is inferred [3–6]. The low energy efficiency in the grid-connected wind energy system is liable for different reasons [PQ-14]. Effect of low power quality such as harmonics, swelling of voltage, dip, instability, fault state, reactive power, and interference problems affect the production of wind power, extra losses, impacting the existence of electrical equipment utilized in the wind power system, direct effect on the grid system with chances of grid breakdown, breach of grid rules, influencing the penetration of wind power. The impact of low power quality problems on the economy of the power system network in the implementation of wind energy is therefore necessary to evaluate.

6 Integration Issues in Grid-Connected Wind Energy

In weak grid areas of the country, the expansion of the wind energy sector has an impact on wind farm dynamics. The essence of electricity output will be influenced because of the increased penetration of the wind energy system into the grid for the aforementioned purposes,

(1) The behavior of the wind energy system is distinct relative to traditional power generation.
(2) In a weak grid, a large number of wind towers are fastened to create a wind farm.
(3) It is not able to support the penetration of strong wind power according to the new weak grid structure.

As per wind grid codes, the power quality problems and dynamics study of wind energy systems are therefore very relevant. The complexities of incorporating wind power into the grid are as follows [31].

7 Effect of Power Quality Issues

7.1 Impact on the Transmission Grid

More VAR usage, reasonable voltage variations, voltage control problems, wind turbine disconnection caused by irregular high or low voltage, wind power output capacity shifts as per wind speed variations, frequency problems, grid extension difficulties, wind power penetration into the grid is restricted, the essence of power injection into the grid within the protection system.

7.2 Impact on the Distribution Grid

The consequences of PQ problems on the distribution grid are voltage dip due to inrush current during system initialization, voltage flicking due to power output fluctuations or tower shadowing effects, and the nature of harmonics with absolute interference of voltages and currents by harmonics.

7.3 Impact of Utility on Wind Farms

The specific conditions for interconnection are the impact of PQ problems on infrastructure and wind farms. Unnecessary wind energy system activation due to steady-state voltage variations on the grid. The Gearbox can be disabled due to abrupt changes in grid voltage. Due to the intrinsic voltage dip on the grid, unwanted tripping of the wind cluster can be possible. Similarly, phase voltage harmonics and unbalancing have had a significant influence on wind energy and utility efficiency.

8 Importance of Custom Power Devices (CPD)

N. Hingorani developed the CPD. It is relevant to the power electronics technology family and offers the delivery system solution for power efficiency. It is common due to the tremendous technical advancement and the ample usage of low-cost high-power switching equipment such as IGBT, GTOs, etc., and recent microcontrollers, microprocessor-based technology. The use of electronic power technologies in the electrical power grid for the benefit of consumers is recognized by custom power. Due to its rapid dynamic reaction, CPD is used as a robust reactive power compensator at the level of delivery and transmission. It is used as a device that integrates among consumer devices and the grid system. It helps to improve the quality of electricity by compensating for electrical disruptions [3–6].

Grid-related wind energy there have been two approaches to resolve the power quality problems either on the grid side or from the wind energy and load side. The first approach is to allow the operation as a load conditioning system under poor quality and another is to be using line conditioning systems to correct for disruptions in power quality. In addition, the supercapacitor, battery, flywheel, and many other storage devices are used to boost PQ problems.

The framework for wind energy requires voltage balance, harmonic load balancing reduction, power factor correction are done by means of a custom power unit shunt linked, i.e., DSTATCOM. The responsive load of wind energy defends against interference by using customized control units linked in sequence, i.e., the dynamic voltage recover. DVR is also used as an active filter in sequence. UPQC is defined as the mixture of active series/shunt filter. The wind and grid-side electricity efficiency problems are addressed by equivalently trying to compensate the supply side as well as the user side with UPQC.

The goals of custom power systems for connected wind energy are as described in the following,

(1) Effective and cheaper implementations of the wind power penetration code of the Indian wind grid.
(2) Choice of introducing new wind power into the grid structure.
(3) Attempting to increase the life of the wind farm and its auxiliary facilities to sustain good power output in the linked wind farm network.
(4) Attempting to reduce the unnecessary load of grid reactive capacity compensation for induction generators to custom power devices with a smart energy storage system.
(5) Improving LVRT and preventing the undesirable interruption of the wind power system linked to the grid during failure.
(6) If adequate load demand is missing during sufficient wind power generation, then the wind power is deposited for a limited period of time in the smart energy storage system and used for internal wind turbine work. This extends equally to the operation of low wind speeds.
(7) Scope for the establishment of a custom power park for potential wind energy as a smart grid technology-focused Power Park.

Grid Code Requirements in Wind Energy System: Power Quality Point of View

9 Introduction

Wind energy is a renewable, inexpensive as well as environmentally friendly means of electric power generation. The major aspect of the wind energy system is the energy transfer from kinetic to mechanical and mechanical to electrical energy.

Due to its accurate variable characteristics, wind power is unstable and thus the variable output is obtained. This raises the complexities of injecting tremendous wind energy into the grid system. For the efficient synchronization dependent interaction of massive wind energy system to the grid requires the consideration of wind turbines link and its specifications to the grid is referred to as grid codes.

From day to day, efficiency and technological development in the power sector together with renewable energy are growing as well as domain knowledge needs to be framed in terms of rules and regulations for the efficient introduction of electric generators into the grid system. It must be assured that the electric power grid is continuously running. This technological guidance is recognized as the grid code that facilitates the enhancement of the efficiency of the electric power system, for the efficient integration of electric generators to the grid.

Although certain general electric grid codes in grid-connected wind energy are not feasible, the exception from such grid code is thus necessary because meeting the

Fig. 9 Power flow of wind energy system

relevant grid codes in the wind farm is economically unfeasible. Therefore, a special grid code is needed for the efficient interaction of wind energy with the grid system. For the effective use of wind power penetration, these grid codes were introduced in different countries. In India, standardized attempts are now being made to introduce wind grid codes efficiently in different regions [8–11, 13].

10 Wind Energy System

Wind energy is an inexpensive and environmentally sustainable way to produce electricity from different renewable sources and therefore has grown rapidly. The remarkable development of wind energy technology in the previous era is illustrated. Due to cheaper characteristics, it became a clear competitor between different power generation services.

11 Basic Concept

Wind mills are available in varying sizes and forms, but they operate the same way. The wind turbine design is normally the same. Similarly, the nacelle begins to spin due to wind, which involves the electric motor and drive train mechanism. It incorporates rotor blades that serve like a turbine. The mechanical power is acquired from the previous case and converted into electrical energy by the generator. The energy transfer of the wind power system at different levels is seen in Fig. 9.

12 Wind Turbine (WT) Technology

The first commercial wind turbine was built in the 1980s and the comprehensive enhancement of the wind turbine in terms of design, power, and efficiency later took place.

Horizontal axis wind turbines are used, with three blades being evenly spaced. The rotor is attached to these blades and power is transmitted via the gearbox to the electric generator. The electric generator and gearbox are included in the Nacelle. The gearbox is missing in a direct drive wind turbine. The injection of wind power into the grid is derived by using a converter. For the efficient running of wind turbines, the wind speed is within 3–25 m/s. Using the new technologies, variable speed wind turbines are used to harness variable wind speed and expand their ability to satisfy grid service requirements. The average installed wind power output was 1.5 MW in 2007. Nowadays, with a rotor diameter of 126 m, the largest wind turbine is 6 MW. Taller and bigger turbines are being used based on new technologies. The new size of wind turbines is 100 times larger than that seen in the 1980s. Similarly, it raises the rotor diameter by eight times. Wind turbines with a power ranging from 1.25–3 MW are commonly utilized on coastline. The weight of blades as well as load of the drive train mechanism now is minimized owing to effective studies within wind energy system. The control system and reliability criteria have now been improved and successful wind turbine–grid integration has been achieved.

13   Basic Mechanism of a WT

In a WT, three blades usually move around a hub. The nacelle is linked to the hub. The nacelle is made up of an electric motor linked by a gearbox. The specific parts of the WT described in Fig. 10 are discussed as below:

Blades of Rotor: A significant element of the wind turbine is the rotor blade. The smaller the diameter, the greater the output power of the wind turbine. Blade design and material are also key variables in a rotor. Wind turbine blades are made of vacuum infusion rubber, epoxy wood, or polyester bonded fiberglass.

Tower: The rotor and nacelle are assisted by it. When erecting a tower, a solid base is installed, acknowledging its sustainability, even throughout situations such as earthquakes. In the tower that descends from the electric motor, a wire is threaded and runs into the converter, which is ultimately attached to the electrical grid.

Nacelle: It is placed on the top of the tower and attached to the blades of the rotor. It consists of a gearbox, electric motor, yaw system, wind vane and anemometer, brake, various sensors, etc. in descending or ascending sequence.

Direct drives and Gearboxes: The gearbox is used to adjust the electric generator's rpm. The new technology is used to drive generators, so more costly gearboxes are not needed.

Brake: The WT blades are halted during heavy storms by a specialized braking mechanism for safety purposes.

Controller: The wind generator module regulates the speed of the blades as well as different electrical components that control the start and stop of the rotor [18, 21, 23, 24].

Electric Generator: Electrical power generation is achieved from it. In the new WT technologies, the sensors are attached to the yaw system, such as wind vanes, anemometers, etc. The direction of wind, the degree of wind speed of WT, is observed by these elements.

Base: On a concrete support base, strong WTs are rendered [2].

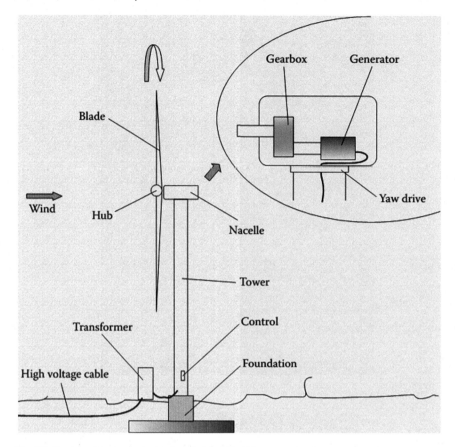

Fig. 10 Basic major components of a wind turbine [2]

14 Vertical Axis WT

The gearbox and the electric generator is centered on the floor and therefore, do not need any tower. The yaw mechanism is not necessary to rotate the rotor against the wind. The downside of the vertical axis wind turbine is that it is not quite inspirational in its performance. In Fig. 11, the Vertical Axis WT is seen.

15 Horizontal Axis WT

The WT shaft is centered on the horizontal axis, and the HAWT travels along the generator's horizontal axis. A traditional horizontal axis wind turbine is represented in Fig. 12.

Fig. 11 Vertical axis wind turbine [2]

Fig. 12 HA WT [31]

16 Installation Location of Wind Turbine

There are two types of wind turbines, depending on construction sites as WT inland and onshore. Availability of steady wind speeds and strong efficiencies are the strengths of offshore WT. Though visual effects, noise, and restrictions are reduced land availability, land use construction, and mountain disputes are often seen as negative of offshore wind turbines.

17 WT Power Output [2]

The power obtained from WT is denoted by symbol Cp and it is called power co-efficient.

$$Pm = Cp\frac{1}{2}\rho A\upsilon^3 = CpPm - W \qquad (1.1)$$

R: radius of blade [m], υ: wind velocity [m/s], Pm: Power extraction, ρ: air density [kg/m³],

Co-efficient of power: Cp, Cp is a function of tip speed ratio and blade pitch angle, where tip speed ratio is defined as

$$\lambda = \frac{R\omega}{\upsilon} \qquad (1.2)$$

ω: angular velocity of WT in rad/s, The ω is calculated from the speed n (Rev/min) by

$$\omega = \frac{2\pi n}{60} \, rad/s \qquad (1.3)$$

$$Cp = \frac{1}{2}(\lambda - 0.022\beta^2 - 5.6)e^{-0.17\lambda} \qquad (1.4)$$

18 Design Speed of Wind Turbine

It is compulsory to choose the correct speed level when designing the WT specified in Fig. 13. Wind output power to achieve the most favorable usage of WT, it is essential to choose the appropriate speed level while designing a wind turbine: cut-in speed; rating speed; and cutoff speed as depicted in Fig. 13 as well as in Fig. 14.

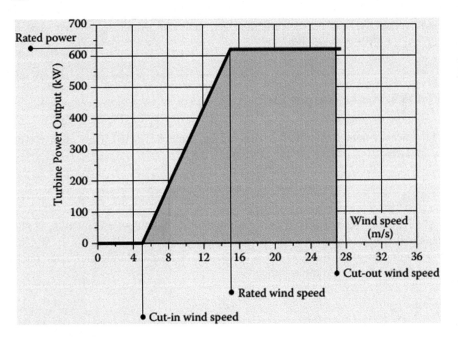

Fig. 13 WT power curve showing three speeds [2]

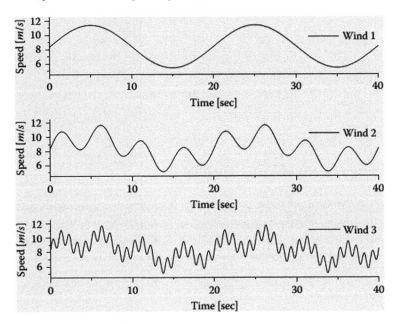

Fig. 14 Wind speed variations

19 Pitch Mechanism

The optimization of power from wind turbine blades is accomplished by pitch modulation and methods of stall control. The optimization of WT power extraction is accomplished by employing the process of pitch and stall function [2].

20 Topologies of Wind Generator [37]

20.1 Introduction

Wind Energy Conversion Systems (WECSs) are a significant feature of the wind energy system and the unit from which electricity is collected and produced is known as the wind generator (WG) [12, 15, 17, 18, 22, 24, 25].

20.2 Fixed-Speed Type a WECS

The aerodynamic rotor driving gearbox and the low-speed–high-speed shaft and asynchronous electrical generator are easy to construct. Figure 15 demonstrates the structure of a Type A WECSS fixed-speed. It requires an asynchronous generator connected through a transformer to the grid. For this case, the running slip fluctuations are normally less than 1%, referred to as a constant speed generator.

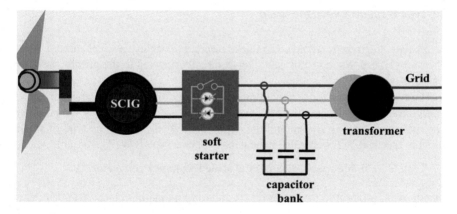

Fig. 15 Fixed-speed type A WECS [37]

The asynchronous induction generator consumes reactive power and thus capacitors are utilized to increase the power factor within every wind turbine. After the start of the induction generator, the minimization of transient state is accomplished by employing a soft starter so in this case the speed is more maintained than the device by the feeding network voltage and drive train. Speed is at its natural velocity. Figure 15 indicates Type A WECS Fixed-Speed.

20.3   Variable Speed WECS

The variable speed wind turbines have the most prevalent type among the built wind mills. The performance of variable WT is the highest relative to fixed-speed WT. In vector WT, maximum power extraction is possible and therefore it is termed as Maximum Power Point Monitoring (MPPT). Consequently, relative to the variable speed WEC, the structure of the fixed-speed WECS is plain. The WECS variable consists of an induction generator (IG) connected directly to the grid through an electronic power interface.

Fluctuations of power due to variable wind direction are controlled by adjusting the speed of the rotor generator. This is done by the use of the converter system for power electronics. Maximum power extraction, reduced mechanical tension, and improved PQ are the merits of variable wind turbines.

The WECS variable speeds are the following.

1.   Type B Limited variable speed
2.   Type C Variable speed with partial scale frequency converter
3.   Type D Variable speed with full-scale frequency converter.

1.   **Type B Limited variable speed**

In this type, the rotor resistance variable is inserted to achieve a small variable rpm. Therefore type B wound rotor generator is directly injected to the grid by variable resistance as well as soft starter.

The reimbursement for reactive power is achieved by using the capacitor bank. This converter, with its costly repair, overcomes the need for slip rings and brushes. The rotor resistance is altered and the slip is controlled. This is the type of small variable speed; B WT. Figure 16 indicates the Speed Type B WECS variable.

2.   **Type C Variable speed with partial scale frequency converter**

This is often referred to as the DFIG (Double Fed Induction Generator). The WT reactive power correction and its incorporation into the grid is carried out by a partial frequency converter. Compared to type B wind turbine, the frequency converter size and large speed regulation is as low as -40 to $+30\%$. For lower frequency converters, this scheme is more desirable and cost-effective. Figure 17 displays the Speed Type C WECSS vector.

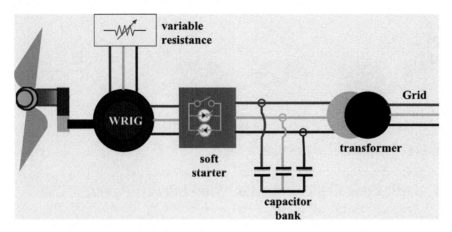

Fig. 1.16 Variable speed type B WECS [37]

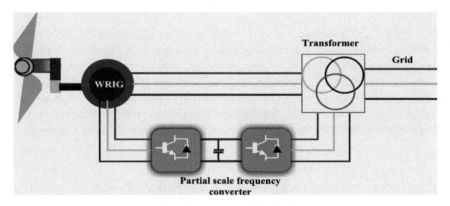

Fig. 1.17 Variable speed type C WECS [37]

3. Type D Variable speed with full-scale frequency converter

This generator, through a full-scale frequency converter, is connected to the grid. The direct drive has a greater diameter multipole electrical generator which is implemented there. The key role of the full-scale frequency converter is to communicate with the grid efficiently and provide efficient compensation for reactive electricity. The scheme uses coil rotor IG and wound rotor synchronous generators or permanent synchronous generators. The scheme of form D is represented in Fig. 18.

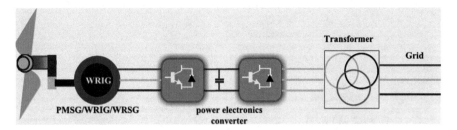

Fig. 1.18 Variable Speed type D WECS [37]

21 Grid Code Requirements in Wind Energy: Power Quality Point of View

21.1 Introduction

Wind energy has variability and regional propagation characteristics. Hence, the injection of WT electricity into the grid is a challenge. It is important to resolve power quality problems in order to successfully incorporate wind power into the grid system [2]. The integration of wind power into the energy grid. The standard technological rules and regulations for the interconnection of enormous WT have had to be explained by different countries and sectors. These formal rules and regulations are termed as grid Codes (GC) for WT interconnection to the grid [12, 15, 17, 22, 24, 25].

21.2 Random Wind Speed Deviation: Normal Operation

There are direct influences on frequency, terminal voltage, electric wind generator output power and they are fluctuating due to spontaneous wind speed deviation. The PQ of WT was thus manipulated. The voltage and frequency of the WT and grid need to be stabilized and some basic control scheme is needed to stabilize both voltage and frequency.

21.3 Transient Stability and Power Quality Problems: Abnormal Operation

The generator speed of the rotor raises and decreases the active power output along with terminal voltage during a fault state in the electric grid network of the attached wind generator. During defective conditions, wind speed is deemed stable. The voltage level goes below 85% of the nominal voltage during such faults. The wind generator is then disconnected from the supply and WT is needed to increase the capability of LVRT/FRT during the fault state.

21.4   Requirements of Grid Connection

New wind power generators are constantly introduced to the electric power grid and it is also important to deal with WT to ensure constant frequency and voltage during steady-state service and grid disruptions. Because of this situation, the grid codes for the injection of WT to the grid are developed or already developed by the different countries and their regular service is stabilized. The Grid Codes are the basic guidelines and regulations for the management of grid reliability for generating stations. The objectives of these grid codes are to confirm the critical implementation of the power quality, reliability, and protection of the electricity system during wind farm expansion.

21.5   Need for Grid Codes

1. Voltage operating range
2. Frequency operating range
3. Power quality
4. Reactive power control
5. Voltage control
6. Communications control
7. Fault ride through capability (FRT).

21.6   Necessity for Indian Wind Grid Code (IWGC) [38]

WT is installed at windy places in remote areas which do not have a solid grid. During low wind power penetration, voltage fluctuations, reactive power usage, and flicker are mainly affected and the overall power system is affected during high wind power penetration. A common practice for wind turbines is established in terms of grid code to address the above problems.

21.7   Grid Behavior of Wind Turbines with Point of Common Coupling (PCC)

For its reactive power compensation, often induction generators are being employed in wind turbines, and the condenser banks are used. The inability to draw reactive power from the electrical grid impacting a possible PCC profile is the result of insufficient capacitance. The synchronous wind turbine generator requires exposure to harmonics and lack of reactive energy issues. Therefore, a standard operating procedure for corresponding grid codes is needed for different types of wind generators utilized in wind turbines.

If the voltage falls below the rated value in PCC, excessive WT retraction is obtained from the grid. Thus, the excessive tricking of the WT is stopped, and WT is in grid linked operating mode, which feeds the power in an appropriate reactive power support during defective state.

21.8   World Situation of Grid Code

Their grid codes, which for wind turbine injection are considerable, are obligatory for different countries such as Denmark, China, Germany, the USA, Spain, Ireland, the Nordic Countries, and Canada. In Asian nations, including India, wind farm grid codes have already been laid up and will be introduced in the days ahead.

21.9   Grid Code Requirements

- **Active Power Control**

In this case it is important to control WT's active output power capability into the power system network. The voltage variations and transient state of WT at start-up and stop-condition of WT can also be accomplished by the maintenance of the frequency.

This will prevent the congestion of the transmission line. If the malfunction has been cleared, it is clean of power rises. Likewise, a wind turbine stays attached to the grid and can solve a malfunction. The current frequency curve shows in Fig. 19.

- **Frequency Requirements**

The energy stability of a power system relative to demand, the reduction in electricity production contributes to frequency reductions below normal value and vice versa. This state can be solved using the main and secondary control system of the synchronous generator. The frequency of the grid can also be influenced by strong

Fig. 19 Power-frequency
curve [39]

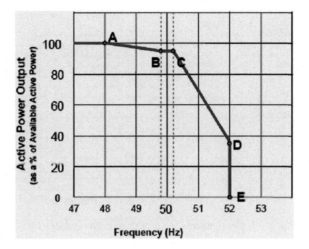

wind energy injection into the grid. During high frequency time the output power
can be smoothed.

- **Voltage and Reactive Power Issues**

Inductor turbines are primarily used in a wind farm, and condenser banks linked to
a wind generator or the grid provide their requirement for reactive power demand.
Reactive power absorption from the grid will lead to lines derated and overheated
and increased losses. This reactive power limitation is not applicable to the wind
farm based on the DFIG synchronous generator. For the wind farm seen in Fig. 20,
a power factor of 0.95–0.95 is typically preserved.

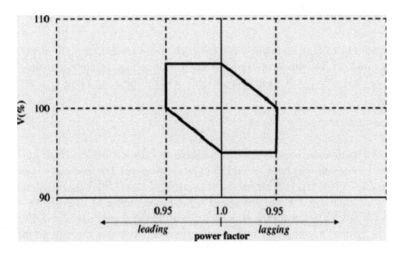

Fig. 20 power factor variation [2]

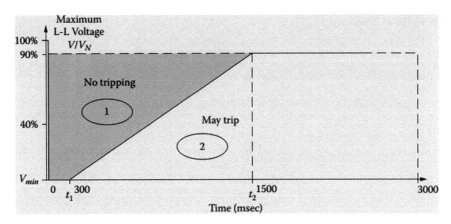

Fig. 21 Typical fault/low voltage ride through capability [2]

- **Fault Ride Through (FRT) Capability**

The wind turbine's ability to remain in the grid connection without tilting of the grid after a certain point after a failure causing the collapse of the strain profile on the computer. FRT time depends on the magnitude of the voltage and the time needed to clear the defect to ensure regular activity at PCC under fault conditions. During breakdown, a voltage failure which in turn raises the power demand of a synchronous generator, which constantly increases, unless a wind generator and its fulfillment from the grid are supplied with a power supply. So the system is becoming unstable. Figure 21 shows the power ride for fault/low voltage.

- **Power Quality Issues**

(1) **Power Quality**

Power issues in voltage, frequency, current fluctuations resulted in equipment loss or malfunction [38]. The numerous problems of power quality depending on the design and associated activity of fixed and variable wind turbines in relation to network activity.

(2) **Harmonics**

The root of harmonics is responsible for variable speed wind turbines. Harmonics are sinusoidal materials that have several integral frequencies. Harmonically injections into variable WT with power electronic converters, such as DFIG, propose harmonic emission calculations for variable speed turbines according to IEC 61400-21 and IEEE STD 1992-519. In compliance with the aforementioned IEEE and IEC standards, the overall harmonic distortion of current and voltages throughout the grid integration of WT is reduced.

(3) **Voltage Imbalance**

The source of harmonics shall be regulated by variable wind turbines. Harmonics are sinusoidal components and has multiple integral frequency. The total voltage and currents of three-phase supply injections in variable WT and splitting the three phase voltage and currents into fractions is just an excess in voltage. These circumstances are found in a wind power system connected to the grid and must therefore be treated as a PQ issue.

(4) **Waveform Distortion**

Through grid injection wind energy in DC offset, harmonics, noise, inter harmonics, sounding, etc., is determined by waveform distortions.

(5) **Sag, Swell, Flicker**

There are brief fluctuations in voltage, which decreased from 0.1 PU to 0.9 PU by 0.5 periods to 1 min.

It is nice to raise the magnitude of voltage by 1.1 Pa to 1.9 PU for 0.5 cycles to 1.0 min. A sudden switching of WT IG using a soft starter, sudden switching of charge, fault etc., and voltage swell occurring due to sudden load switching, etc. Voltage slope and swell in a wind system is created.

During the extensive integration of the wind energy grid, voltage flickering occurs. According to IEEE standards and IEC, we need to track and manage the above PQ problems in the grid-connected mode of operation of the wind energy system in full accordance with the grid codes.

22 Summary

The present and prospective energy scenario was discussed in relation with other countries regarding the availability and usage of different forms of green and traditional energy resources in India. The need and relevance of wind power use have been addressed. The expansion of the power sector, obstacles to the production of wind energy, the significance of PQ in grid-connected WT, integration problems, the importance of CPD in WT were addressed. According to the Government of India's "moving from megawatt (MW) to giga-watt (GW)" slogan, there is increasingly growing potential for the use of the wind energy infrastructure.

In this chapter, grid codes, WECS, basic WT components, WT types, various WG and WT topologies are listed. At the time of wind power injection to the grid, the grid-connected wind farm with power quality problems is stated.

References

1. Ackermann T (2005) Wind power in power systems. Wiley, Chichester, England
2. Ali MH (2012) Wind energy systems. CRC Press, Taylor & Francis Group, London
3. Bhadane KV, Ballal MS, Moharil RM (2012) Investigation for causes of poor power quality in grid connected wind energy—a review. Asia-Pacific Power Energy Eng Conf 1–6
4. Bhadane KV, Ballal MS, Moharil RM, Suryawanshi HM (2016) A control and protection model for the distributed generation and energy storage systems in microgrids. J Power Electron 16(2):748–759
5. Bhadane KV, Ballal MS, Moharil RM, Suryawanshi HM (2014) Wavelet transform based power quality analysis of grid connected wind farm—an investigation of power quality disturbances. In: 2014 international conference on advances in electrical engineering (ICAEE), pp 1–6
6. Bhadane KV, Ballal MS, Moharil RM, Suryawanshi HM (2015) Enhancement of distributed generation by using custom power device. J Electron Sci Technol 13(3):246–254
7. Burton T, Sharpe D, Jenkin N, Bossanyi E (2001) Wind energy handbook. John Wiley & sons Ltd., Chichester
8. Caramia P, Carpinelli G, Verde P (2009) Power quality indices in liberalized markets, 1st edn. Wiley, Chichester
9. Carlos V, Sergio M, Francisco B (2005) Large scale integration of wind energy into power systems. Electr Power Qual Util Mag 1(1):1–6
10. Chen Z, Spooner E (2001) Grid power quality with variable speed wind turbines. IEEE Trans Energy Convers 16(2):148–154
11. Divya KC, Nagendra Rao PS (2006) Models for wind turbine generating systems and their application in load flow studies. Electr Power Syst Res 76:844–856
12. Dusonchet L, Telaretti E (2011) Effects of electrical and mechanical parameters on the transient voltage stability of a fixed speed wind turbine. Electr Power Sys Res 81:1308–1316
13. Ewing RA (2003) Power with nature, 1st edn. Pixy jack Press, USA
14. Ghosh A, Ledwich G (2002) Power quality enhancement using custom power devices. Springer Science + Business Media, LLC, 1st Edition, published by Kluwer Academic Publishers, New York
15. Grillo S, Massucco S, Morini A, Pitto A, Silvestro F (2010) Micro turbine control modeling to investigate the effects of distributed generation in electric energy networks. IEEE Sys J 4(3):303–312
16. Juan MC, Leopoldo GF, Narciso MA (2006) Power electronic systems for the grid integration of renewable energy sources: a survey. IEEE Trans Ind Electron 53(4):1002–1016
17. Kabouris J, Kanellos (2009) FD Impacts of large scale wind penetration on energy supply industry. Energies 2:1031–1041
18. Li H, Chen Z (2008) Overview of different wind generator systems and their comparisons. IET Renew Power Gener 2(2):123–138
19. Manwell JF, Mcgowan JG, Rogers AL (2002) Wind energy explained: theory, design and application. John Wiley & sons Ltd., Chichester
20. Mehta VK (2013) Principles of power system, 2nd edn. S Chand Publication, New Delhi
21. Prasad H, Bhadane KV (2020) Recent control and integration issue of distributed energy resources in smart microgrid: a review. Int Conf Innov Trends Adv Eng Tecnol
22. Skretas SB, Papadopoulos DP (2009) Efficient design and simulation of an expandable hybrid (wind–photovoltaic) power system with MPPT and inverter input voltage regulation features in compliance with electric grid requirements. Electr Power Sys Res 79:1271–1285
23. Slootweg JG, de Haan SWH, Polinder H, Kling WL (2003) General model for representing variable speed wind turbines in power system dynamics simulations. IEEE Trans Power Syst 18(1):144–151
24. Tsili M, Papathanassiou S (2009) A review of grid code technical requirements for wind farms. IET Renew Power Gener 3(3):308–332
25. Wang L, Chen LY (2011) Reduction of power fluctuations of a large-scale grid-connected offshore wind farm using a variable frequency transformer. IEEE Trans Sustain Energy 2(3):226–234

References from Website

26. http://www.nrel.gov/wind
27. http://www.cea.nic.in/reports/monthly/inst_capacity/may15.pdf
28. http://www.cea.nic.in/reports/monthly/inst_capacity/jan15.pdf
29. http://mnre.gov.in/file-manager/akshay-urja/january-february-2015/EN/files/basic-html/page5.html
30. http://orionrenewables.com/wind/why-wind-energy
31. http://ewh.ieee.org/r4/milwaukee/pes/wind.html
32. http://www.cwet.res.in/AdThirunoorthy/CChellamuthu/RD-RD-190-10/March2014
33. www.mospi.gov.in
34. www.worldenergy.org
35. www.mnre.gov.in/strategicplan2011
36. www.niwe.res.in
37. www.ethesis.nitrki.ac.in/satishchoudhury
38. www.cwet/rajeshtalyar/Indianwindgridcodereport
39. www.aemo.com.au/windintegration/internationalexperience/reviewofgridcodes2011
40. www.onlineelibrary.wiely.com/gabrielemichalke/varia.s.windturbinethesis
41. www.mathworks.com/themathworks,Incsimpowersystemsrreference2007

New Software for Matching Between Wind Sites and Wind Turbines

Ali M. Eltamaly

Abstract Wind energy system is becoming a mature technology for generating electric energy from the wind. The design of the wind energy system should take into consideration the matching between the wind speed site characteristics and the performance characteristics of the wind turbine. This chapter is introduced to perform the matching process between the site and the wind turbines (WTs) for minimum cost and highest reliability. An accurate matching methodology for pairing between site and WTs has been introduced. The pairing methodology is designed in Matlab code to perform this study. The input data for 32 Saudi Arabia sites and 140 market available WTs have been selected to validate the right operation of the new proposed computer program. This program will select the best site and the most suitable WT for this site based on techno-economical methodology. This program can help researchers, designers, experts, and decision-makers to select the best site among many sites and the best WT for each site. The results obtained from this site show a substantial reduction in cost when the best site is selected as the most suitable WT for this site.

1 Introduction

Wind turbines (WTs) have different technologies and different performance characteristics. At the same time, wind speed characteristics of sites are different from one site to another. Where one WT can generate high energy in some sites, meanwhile it may generate lower energy in some other sites because of the correlation between the wind turbine performance parameters and site parameters. To get the minimum cost of energy the suitable WT should be selected for minimum cost. This operation is called a pairing between the site and the WT. Based on this speed the suitable WT

A. M. Eltamaly (✉)
Sustainable Energy Technologies Center, King Saud University, Riyadh 11421, Saudi Arabia
e-mail: eltamaly@ksu.edu.sa

Electrical Engineering Department, Mansoura University, Mansoura, Egypt

K.A. CARE Energy Research and Innovation Center, Riyadh 11451, Saudi Arabia

is operating at high efficiency in this site which can generate higher energy in this site which can be translated to a substantial reduction in the cost of energy. The most important parameters that characterize the operation of the WTs are the cut-in, rated, and furling wind speeds which highly affect the performance of the wind energy systems. The parameters of wind speeds sites that can characterize the wind speed profile of this site are the Weibull scale, c and shape, k parameters, and average wind speed. These two parameters can describe the wind speed profile better than many other statistical models such as gamma and Rayleigh models [1–12].

One of the simplest methodologies used to pair between sites and WT is done by estimating the average power output from the WT by using it in a certain site. Reference [1] performs direct numerical integrations for the output power corresponding to each wind speed to determine the average output power from WT. Despite the simplicity of this method, it needs a computer program to perform numerical integration [1]. On the other hand, reference [4] introduced some modifications to the previous method [1] by introducing formula to determine the average output power as a function of WT and site parameters.

Reference [4] proposes a technique to choose the best option from some of WT and sites. The wind speed data of wind sites and the data for WTs will be used to predict the capacity factors for each combination and the maximum value of capacity factor will be selected as the best option, and the site and WT were selected.

The Weibull parameters have been used for determining the wind profile for 32 sites in many countries around the world from data collected for several years. Also, the specifications of 20 different WT types were collected. The Weibull distributions of all 32 sites can be categorized into four types, type A to Type D. These sites are put in order from the lowest to highest average wind speeds where type A represents the sites with the lowest wind speed and Type D represents the highest one. The capacity factor, C_F is calculated for each WT–site pair, type A sites have very bad C_F indices because they do not have enough wind speed to reach the threshold of cut-in speeds. And type D has the highest capacity factor. It is concluded from this study that the WT has high cut-in wind speed and the high-rated wind speed will have a lower capacity factor and it represents an unsuitable candidate to be used in lower wind speed sites [4]. This methodology gets the final results based on the technical constraints and neglects the economic constraints. Due to the variation of the price for each kW of different WT types, the decision may not give us the best price for generated kWh. On the other hand, the yearly generated energy, and the cost of kWh produced were not estimated in these references [4, 5] and so this reduces the reliance upon in the assessment process.

Another study uses the Geographic Information System (GIS) to determine an analytic framework to evaluate site suitability for WTs depending on its wind speed and to select the best site based on an economical point of view. This methodology uses rule-based spatial analysis to determine suitable sites for WT placement. The selection strategy includes the wind energy components such as wind resources, and site obstacles, and terrain and some other environmental factors and human impact factors. The highest consideration is provided for the physical factors such as the placement of central loads to choose the best placement of the wind turbines taking the other factors such as the environmental and human impact factors in less consideration. The location of the intended wind energy project has been shown based on each separate factor and also based on the combination of more than one factor together [5]. This technique chooses sites depending on the highest average wind speed and overlooking the WT performance parameters which are the main drawback of this technique [5]. This technique [5] omitted assessment in economic terms, as in the previous methodologies [1, 4].

The capacity factor is used in many studies to pair between the site and the WT which is introduced in [4, 5]. The main drawback of this study is its building the pairing decision based on technical factors without giving the cost analysis any weight which is not enough to take this decision where the assessment steps should be based on the technical and economic factors. Taking the technical and economic factors are introduced in many studies in many sites in Saudi Arabia and Egypt [9, 13–15]. Another strategy used the value of money method to determine the present value of costs, *PVC*, of generated energy for the complete year. The *PVC* is used to estimate the cost generated energy by dividing the *PVC* by the total generated energy produced through the whole lifetime of the project [9, 13–15]. Another study used the monthly average wind speed of seven measurement stations located at the east coast of Egypt and two WT for pairing sites and WT [9].

The rest sections of this chapter are showing the models of WT in Sect. 2. The detailed discretion of the proposed computer program is introduced in Sect. 3. The results extracted from the 32 sites and the 140 WTs under study are introduced in Sect. 4. The last section is showing the conclusions extracted from the new proposed computer program.

2 Modeling of Wind Energy System

The mechanical energy extracted from wind is done by extracting the kinetic energy from moving winds. The wind power in the wind turbine depends on the wind speed, the cross-sectional area of the wind stream, and density of air at the site as shown in (1). It is clear from this equation that the power is directly proportional to the power three of the wind speed which means that the wind power is considerably affected by the wind speed. The height of the WT has a great effect on the wind speed and the generated energy from the WT. Most metrological stations are located at a level called the measurement height, h_g. Most of the wind speed measurement stations are located at 10 m above the ground level. The relationship governing the wind speed at any height, h is shown in Eq. (2) [2]. The WT cannot extract the whole generated power in the wind. The highest theoretical value that can be extracted from wind is 16/27 (0.593) which is called the Betz coefficient. Moreover, the wind turbine cannot start spinning at very low wind speed due to the friction and inertia weight of rotating components. Instead, it can spin and start to produce power at a certain wind speed called cut-in. At this speed, the WT can start to produce electric power. The cut-in wind speed differs from WT to another where its value is between 2.5 and

3.5 in most WTs. The good WTs have low cut-in wind speeds. The WT generates more power when the wind speeds exceed the cut-in wind speed until the wind speed reaches the rated wind speed of WT, u_r. Beyond the rated wind speed, the power will hold constant when the wind speed exceeds the rated wind speed to protect the components of wind turbines from overloaded conditions. The relationship between the generated power from WT against wind speed is different from one WT to another, but most of the WT has the relation as shown in Eq. (3). The generated power should stop once the wind speed reaches a certain high wind speed called cut-out or furling wind speed, u_f to protect the WT from damage due to very high wind speeds. The relation between the power in the wind and generated power from WT against wind speed is shown in Fig. 1. This relation shows the importance of wind speed on the generated power from WTs. So, it is recommended to choose sites with high average wind speed to extract more power. Also, it is recommended to choose WT with rated wind speed parameters suitable for this site. Based on the importance of wind speed on the generated power from WTs, many studies were introduced in the literature to select the best possible site from many sites [10, 14–51] which are very important before start installation of the wind energy projects. The generated power from WT is shown in (1).

$$P_w = \frac{1}{2}\rho A_t u^3 \tag{1}$$

where ρ is the air density.

A_t is the cross-sectional area of wind parcel, m

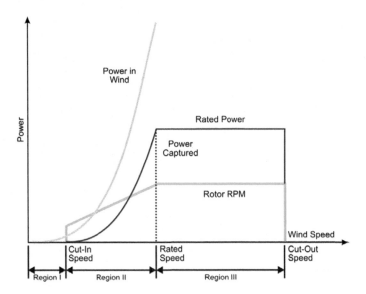

Fig. 1 Actual WT output power with the wind speed

u is the wind speed m/s

$$u(h) = u(hg) * \left(\frac{h}{hg}\right)^{\alpha} \tag{2}$$

where h: The height that the wind speed is needed to be determined above the ground level, m.

hg: The height of measuring the wind speed, m.

α: The power-law exponent, which depends on the roughness of the ground surface, its average value, is (1/7) [2].

$$P_W(u) = \begin{cases} 0 & u \leq u_C \\ P_r * \frac{u-u_C}{u_r-u_C}, & u_C \leq u \leq u_r \\ P_r & u \geq u_r \\ 0 & u \geq u_f \end{cases} \tag{3}$$

The above equation is used to determine the generated power from the WT based on the wind speed parameters of the WT.

The wind speed is changing all the time, and for this reason, it should be modeled using statistical techniques. Weibull distribution is one of the most famous techniques used to model the speed of the wind as shown in the following equations;

The average wind speed u_{av} of measured wind speeds, u_i is shown in Eq. (4)

$$u_{av} = \frac{1}{n}\sum_{i=1}^{n} u_i \tag{4}$$

The Weibull density function, $f(u)$ that can represent the wind speed variation is shown in (5) where this function represents the frequency of speeds in the measurement data. This function has two parameters called the scale parameter, c (m/s), and shape parameter, k. The shape of distributed function for different values of shape factor, k at scale factor $c = 1$ is shown in Fig. 2. It is clear from this figure that the function is getting narrow as the shape factor is getting high value and vice versa.

$$f(u) = \frac{k}{c}\left(\frac{u}{c}\right)^{k-1} \exp\left(-\left(\frac{u}{c}\right)^k\right), \quad (k > 0, \ u > 0, \ c > 1) \tag{5}$$

There are many ways to determine c and k parameters as shown in the following equations. The empirical relationship between c and u_r as shown in Eq. (6) can give a good estimate when the shape factor is between 1.5 and 3 [2].

$$c = 1.12\bar{u} \quad (1.5 \leq k \leq 3.0) \tag{6}$$

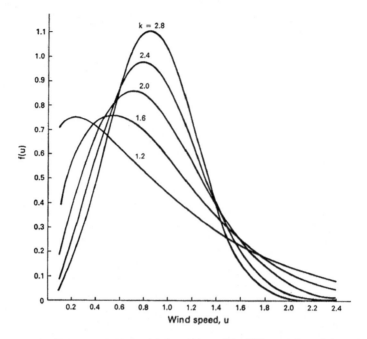

Fig. 2 The density function $f(u)$ along with the wind speed for different values of shape parameters at $c = 1$ [2]

Similar relation (7) can be used to estimate the shape parameter, k against the average wind speed, and variance of wind speeds [2].

$$k = \left(\frac{\sigma}{\bar{u}}\right)^{-1.086} \tag{7}$$

The standard deviation can be determined from (8) and the variance can be determined from this equation by taking the root of standard deviation.

$$\sigma^2 = c^2\left[\Gamma\left(1 + \frac{2}{k}\right) - \Gamma^2\left(1 + \frac{1}{k}\right)\right] = (u)^2\left[\frac{\Gamma(1 + 2/k)}{\Gamma^2(1 + 1/k)} - 1\right] \tag{8}$$

Another Eq. (9) can be used to determine the relation between average wind speed and scale and shape parameters which can give an accurate estimation for these parameters in the range of shape parameters between 1 and 10 ($1 \leq k \leq 10$).

$$c = \frac{u}{\Gamma(1 + 1/k)} \tag{9}$$

Another statistical technique can be used to determine the scale and shape parameters as shown in (10) [52, 53].

$$k = a$$
$$c = \exp(-b/k) \tag{10}$$

where:

$$a = \frac{\sum_{i=1}^{w} x_i y_i - \frac{\sum_{i=1}^{w} x_i \sum_{i=1}^{w} y_i}{w}}{\sum_{i=1}^{w} x_i^2 - \frac{\left(\sum_{i=1}^{w} x_i\right)^2}{w}} = \frac{\sum_{i=1}^{w} (x_i - \bar{x}) \sum_{i=1}^{w} (y_i - \bar{y})}{\sum_{i=1}^{w} (x_i - \bar{x})^2} \tag{11}$$

$$b = \bar{y}_i - a\bar{x}_i = \frac{1}{w} \sum_{i=1}^{w} y_i - \frac{a}{w} \sum_{i=1}^{w} x_i \tag{12}$$

$$\text{And} \quad y_i = \ln(-\ln(1 - F(u_i))),$$
$$x_i = \ln(u_i) \tag{13}$$

The value of the capacity factor, CF using the site and WT parameters is shown in (14).

$$C_F = \frac{\exp\left[-(u_C/c)^k\right] - \exp\left[-(u_r/c)^k\right]}{(u_r/c)^k - (u_C/c)^k} - \exp\left[-(u_F/c)^k\right] \tag{14}$$

$$P_{W,av} = C_F * P_r \tag{15}$$

The average number of WTs, ANWT is determined by dividing the average load value $P_{LW,av}$ by the average power generated from WT as shown in Eq. (16).

$$ANWT = \frac{P_{LW,av}}{P_{W,av}} \tag{16}$$

3 The Proposed Computer Program Implementation

The first step of building a wind energy system in any country is to select the windy sites with high average wind speed and near to the load centers. These sites should be subject to more studies to select the best one and the best WT that can produce the required load with the lowest cost. The WT should be paired with the sites because the WTs can work economically in one site but it may not be suitable for another site. For this reason, the sites and the WTs data should be collected to select the best site and the best WT suitable for this site for the minimum cost of energy. For this reason, 32 sites were collected in many places in Saudi Arabia and 140 market available WTs were also selected to be studied. The detailed performance data of 10 of the 140 WTs are shown in Table 1. An efficient computer program is designed to

Table 1 Sample of the WT data used in the computer program

ID	Manufacturer	Model	Offshore	Power	Diameter	Hub min height	Hub max height	Min wind speed	Nominal wind speed	Max wind speed	Gear box	Output voltage	Power control
No	Company Name	Model Number	Installation Place	kW	m	m,	m	m/s,	m/s,	m/s,		V	
1	Enercon	E33/330	No	330	33.4	37	50	3	13	34	no	400	Pitch
2	Enercon	E44/900	No	900	44	45	65	3	13	34	no	400	Pitch
3	Enercon	E48/800	No	800	48	50	76	3	13	34	no	400	Pitch
4	Enercon	E53/800	No	800	52.9	60	75	2	13	34	no	400	Pitch
5	Enercon	E70/2300	No	2300	71	57	98	2	15	34	no	400	Pitch
6	Enercon	E82/2000	No	2000	82	78	138	2	13	34	no	400	Pitch
7	Nordex	N90/2500	Yes	2500	90	70	80	3	13	25	yes	690	Pitch
8	Nordex	N90/2300	Yes	2300	90	80	105	3	13	25	yes	690	Pitch
9	GE Energy	4.0-110	Yes	4000	110	80	110	3	14	25	no	690	Pitch
10	Nordex	S77	No	1500	77	61.5	100	3.5	12.5	25	yes	690	Pitch

perform the matching steps of the sites and the WTs. Many useful results like the best site and best WT for this site and other sites and the cost of generated energy from all sites with different WTs can be extracted from this computer program. Due to the generic nature of this computer program, it can be used in any place of the world and it can help decision-makers to select the best option to install the wind energy system and to predict the benefits out of the installation of these projects. The new proposed computer program has more flexibility and more accurate results than the market available commercial design programs like the HOMER or the RetScreen. The block diagram showing the logic of this computer program is shown in Fig. 3. This computer program is having five subroutines having flowcharts showing their logic as shown in Figs. 4, 5, 6, 7 and 8. These subroutines are listed in the following points and they will be discussed in the following sections.

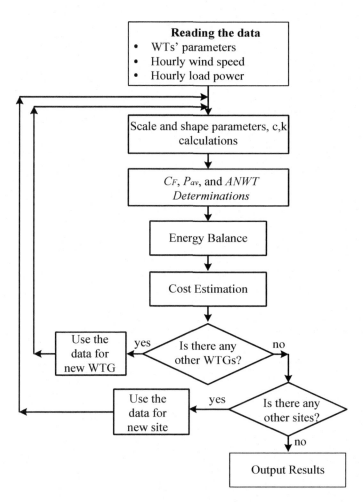

Fig. 3 The main logic of the computer program

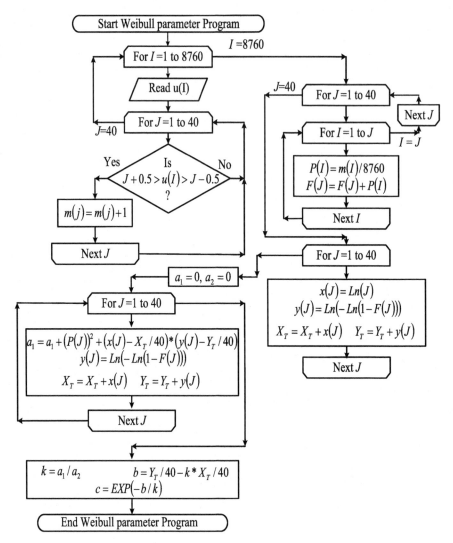

Fig. 4 The flowchart used to determine the Weibull parameters (c, k) as discussed in the first subroutine

- Weibull Parameters Determinations.
- Average Power Generated from WES.
- Energy Balance.
- Outputs from Economic Analyses.
- Modified Economic Model.

The proposed computer program is built in Matlab. The data used in the program are the hourly wind speed with their measuring height, the hourly load values, and the

Fig. 5 The flowchart of the second subroutine for calculating the capacity factor and average number of WTs

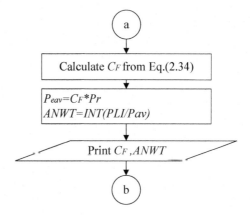

WTs data and their performance parameters and cost. The optimization procedures of this computer program will be performed to choose the best site and best WT for this site and the cost of generated kWh. All these data are saved in excel files which will be connected to the computer program to handle the design and optimization steps. The proposed program is a built-in generic form to be used with any number of sites and any number of WTs which make it suitable to be used in any place in the world. The following sections show the details of the main program, its input/output, and the detailed description of the subroutines.

3.1 Input Data

The detailed descriptions of the data introduced to the computer program are shown in the following points:

1. The detailed wind speed data of the 32 sites are introduced to the computer program. These data are the hourly wind speeds for 32 sites in Saudi Arabia and the height of the measurement stations. The computer program can handle an unlimited number of sites and an unlimited number of WTs. These data were selected for several years in many places in Saudi Arabia to have high confidence in the results obtained from this program. The summarized data for these selected 32 sites are shown in Table 2.
2. The performance data for the market available 140 WTs are introduced to the new proposed computer program. These performance characteristics of WTs are the rated power, hub height, diameter of swept area, efficiency of the WTs components, cut-in speed, rated speed, cut-out (furling) speed, and the price of WTs. The selected WT types are chosen in different technologies, sizes, and manufacturers. Most of the WTs are selected in size greater than 200 kW.
3. The detailed load characteristics should be introduced to the program; the loads are introduced in hourly fashions. These data are introduced in this chapter.

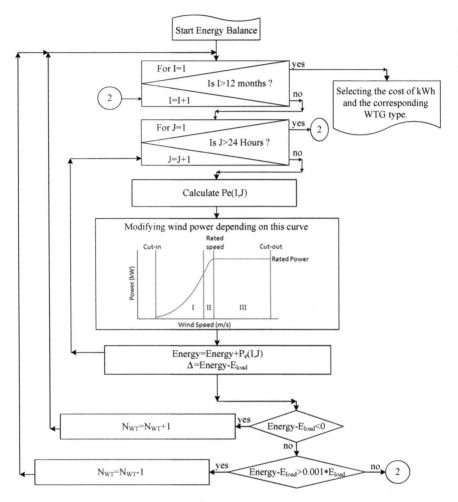

Fig. 6 The flowchart of energy balance subroutine

4. The hourly load profile should be introduced to the computer program. The load data used in this program are collected from actual data of remote communities of one of the villages located at the north of Saudi Arabia called Addfa in the Al-Jouf region province. The hourly data of these loads are introduced to the proposed program. The monthly maximum, minimum, and average loads are shown in Fig. 9.

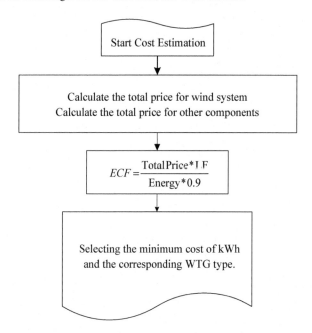

Fig. 7 The flowchart of the energy cost of kWh for each combination and minimum cost of kWh generated and the corresponding WT type

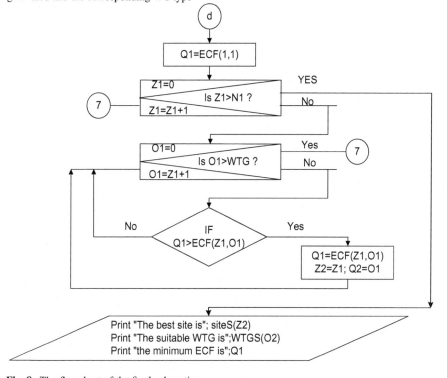

Fig. 8 The flowchart of the final subroutine

Table 2 Summarized data for the selected 32 sites

Site#	Site Name	Longitude	Latitudes	u_{av}	c	k
1	Dammam	50.16	26.4	4.4	4.624301	1.754418
2	Kfia	50.15	26.3	4.41	4.509051	2.755985
3	Dhahran	50.12	26.32	4.31	4.423661	2.920424
4	Al-Ahsa	49.6	25.34	3.7	3.741851	1.877328
5	Sharorah	47.1	17.3	3.31	3.332941	2.556731
6	Riyadh	46.5	24.4	2.95	2.940099	1.939081
7	Qaisumah	46.1	28.18	3.71	3.772909	2.141096
8	Hafer-AlBaten	45.7	28.4	3.33	3.486257	1.609102
9	Asulil	45.6	20.47	3.80	3.978311	2.202145
10	Wadi Al-Dawasser	44.15	20.3	3.5	3.543813	2.193621
11	Najran	44.15	17.35	2.12	2.021776	2.101696
12	Gassim	43.9	26.17	2.91	2.902935	2.116965
13	Rafha	43.5	29.6	3.89	3.973186	2.295766
14	Gizan	42.55	16.9	3.15	3.09359	3.765881
15	Abha	42.51	18.15	3.13	3.088384	2.409208
16	Khamis Mushait	42.43	18.25	3.02	3.000436	2.511597
17	Bisha	42.4	20.25	2.45	2.399268	2.280234
18	Douhlom	42.17	22.72	4.32	4.434125	1.742083
19	Hail	41.6	27.35	3.98	3.48484	2.551401
20	Al-Baha	41.47	20.17	3.46	3.701758	2.495482
21	Taif	40.42	21.27	3.67	3.834777	1.662116
22	Skaka	40.17	29.9	3.82	3.963052	2.220434
23	Al-Jouf	40.1	29.8	3.9	3.678546	2.369471
24	Arar	41.02	30.95	3.63	1.500668	2.097325
25	Makkah	39.84	21.42	1.64	3.011625	2.841229
26	Madinah	39.6	24.45	3.05	3.682481	2.909928
27	Jeddah KAIA	39.2	21.7	3.65	4.263647	2.738051
28	Turaif	38.7	31.7	4.17	3.839341	2.640686
29	Yanbou	38.1	24.1	3.77	4.365416	2.050805
30	Guriat	37.3	31.4	4.24	4.563931	3.43555
31	Wejh	36.75	26.33	4.52	2.587132	1.356426
32	Tabouk	36.57	28.4	2.72	4.713015	3.032604

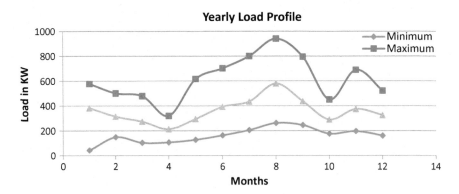

Fig. 9 Monthly minimum, maximum and average load profile for Addfa

3.2 Proposed Computer Program Steps

The block diagram shows the main logic of the proposed computer program as shown in Fig. 3. The part that takes the input data listed above will be introduced to the main computer program. These data should be introduced to the parts of the computer program, and it will be fed to the different subroutines. The main program is controlling the logic of the five subroutines to make different calculations and optimizations. These subroutines perform different parts of the calculations and optimizations and the logic of these subroutines are shown in the following sections.

3.2.1 The Weibull Parameters Determinations "First Subroutine"

This is the first subroutine and it is used to determine the Weibull parameters (scale and shape parameters, c, k). The flowchart used to implement the logic of this subroutine is shown in Fig. 4. The hourly wind speed at the measurement height will be modified from Eq. (2) to the hub height of the WTs. The calculations of Weibull parameters are done using the Eqs. (4)–(14). The results obtained from this subroutine are compared from the same logic used in the excel file and the results obtained from market available programs like the HOMER and the RetScreen software. The results from this subroutine are very near to the results obtained from other programs that give confidence to the obtained results.

3.3 Average Number of WTs and Capacity Factor Determination "Second Subroutine"

The second subroutine is using the results obtained from the first subroutine to this subroutine to calculate the capacity factor, C_F as shown in Eq. (14), the average power of the WT as shown in Eq. (15), and the average number of WT, $ANWT$ as shown in (16). It is worth to be noted that the site and WT having the highest number of capacity factor gives an indication for the best option of pairing between the wind turbine and site based on technical perspective meanwhile the cost calculations give us the final and techno-economical results. The average number of wind turbine calculated from this subroutine will be used as a start value to the third subroutine (Energy balance subroutine) to determine the real number of WTs needed to cover the hourly load within completely.

3.4 Energy Balance "Third Subroutine"

The subroutine is used to check the adequacy of the hourly generated power from the WES to feed the loads. The generated power from the WES is shown in Eq. (17). The overall efficiency of the WT is shown in Eq. (18). If the case of the load requirement is greater than the generated power from the WES, the deficit power will be supplied from the electric utility. Meanwhile, in the case of the hourly generated power from the wind energy system is greater than the power required by the loads, the extra power will be transmitted to the electric energy. The energy balance subroutine should assure that the energy fed to the electric utility is equal to the energy received from the utility grid. This logic is modeled in the Eqs. (19), (20), and (21).

The output power from WES is given by:

$$P_e = \frac{1}{2} \rho * A * u^3 * N_t * \eta_0 \tag{17}$$

$$\eta_o = C_p \, \eta_m \eta_g \tag{18}$$

$$\text{If } P_e > P_L, \text{ Then } P_T = P_e - P_L, \text{ and } P_f = 0 \tag{19}$$

$$\text{If } P_e < P_L, \text{ Then } P_F = P_L - P_e \text{ and } P_T = 0 \tag{20}$$

For energy balance the following conditions must be satisfied;

$$\sum_{i=1}^{8760} P_e(i) - \sum_{i=1}^{8760} P_L(i) = 0 \quad \text{and} \quad \sum_{i=1}^{8760} P_T(i) = \sum_{i=1}^{8760} P_F(i) = 0 \tag{21}$$

3.5 Economic Analyses "Fourth Subroutine"

3.5.1 Overview of Outputs

Four main outputs are applicable to the economic analysis of prospective wind farms. These include the NPV, the present cost of energy, and the time to recover capital. This subroutine will handle the calculations of the cost of energy for the WTs and sites under study. The flowchart of this subroutine is shown in Fig. 7. Several works are already presented as a first step to harness this alternated energy source and to use this to generate electricity at a large and economical scale. A detailed technique is used to calculate the generated kWh from WES. All the details of the WES are considered in this methodology to a precise estimation for the cost of kWh generated from the WES. This methodology is applied for all sites and all wind turbines. The site and WT corresponding to the lowest cost will be selected as the best option. Also, the lowest cost associated with each site can be used to select the best WT for this site. The results obtained from the cost analysis methodology can produce many results as will be introduced in the following sections.

The price of the kW rated power of the WT is used as \$700/kW. Equation (22) is used to determine the total price of WTs by multiplying the price per kW by the number of WTs and the rated power of the WT. The price of the microprocessor (TPMIC), the price of main substation (TPMS), the price of the modem for remote control in central control station (TPCCS), and the price of the transmission line (TPTL) are 2.3 \$/kW, 10.4 \$/kW, 4.16 \$/kW, and 1.3 \$/kW, respectively, and the total values of these items are shown in Eqs. (22)–(26).

$$TPWT = \$700 * NWT * P_r \tag{22}$$

$$TPMIC = \$2.3 * NWT * P_r \tag{23}$$

$$TPMS = \$10.4 * NWT * P_r \tag{24}$$

$$TPCCS = \$4.16 * NWT * P_r \tag{25}$$

$$TPTL = \$1.3 * NWT * P_r \tag{26}$$

where:

TPWT   Total price of WTs.
TPMIC   Total price of microcontrollers.
TPMS   Total price of the main substation.
TPCCS   Total price of remote control in the central control station.
TPTL   Total price of the transmission lines.

The maintenance and operation cost is taken as 10% of the total cost of the WES as shown in (27). The energy cost figure (*ECF*) can be obtained from using the total cost of the WES as shown in (28) and divide it by the yearly generated energy multiplied by the levelization factor as shown in (29). The total price of WES can be calculated from (27) and (28) and the *ECF* can be determined from (30).

$$Total\ Price = 1.1 * (TPWT + TPMIC + TPMS + TPRC + TPCCS + TPTL) \tag{27}$$

$$Total\ Price = 1.1 * (718.16 * NWT * Pr) \tag{28}$$

$$ECF = \frac{Total\ Price\ * LF}{YE * 0.9} \tag{29}$$

where:

LF is the levelization factor.

The levelization factor is 0.177 based on a 12% interest rate and 10 years recovery time.

3.5.2 Modified Economic Model

The design of wind energy systems should be based on techno-economical methodology where the cost of the generated energy is the most important issue. The WT may work efficiently in economical in some sites, but it may not work in some other sites. For this reason, the selection of the WT and the site should be based on the cost of energy. To obtain the highest available energy, the WT can have enormous blades and wide swept area by these blades; however, this not economical because the wide-area required a huge and strong tower and high technology for manufacturing the blades and other components of the WTs and all these come at a cost. Also, the most efficient turbine may not be the cheapest to produce energy because it may be suitable for one site and is not suitable for another one. To compare the cost of energy produced by a turbine, some economic analyses need to be performed.

Different economical methodologies have been introduced in the literature to calculate the expected price of generated energy from the WES. Some methodologies used the NPV to perform the calculations of the cost of kWh [54, 55]. These methodologies determined the Levelized Cost of Energy, *LCE* to calculate the cost of energy as shown in (30) [56]. The value of LEC can be determined by dividing the total present value multiplied by the capital recovery factor, *CRF*.

$$LCE = \frac{TPV * CRF}{AE} \tag{30}$$

where TPV is the total present value of the wind energy system, *AE* is the yearly loads connected to the WES. The *CRF* is the capital recovery factor and can be obtained from (31).

$$CRF = \frac{r(1+r)^T}{(1+r)^T - 1} \tag{31}$$

Equation (32) is used to determine the TPV by adding the whole cost and subtract the components' salvage values at the beginning of the operation of the WES.

$$TPV = IC + RC + OMC - PSV \tag{32}$$

where *IC* is the initial capital cost of the whole system, *RC* is the replacement cost, *OMC* is the operation and maintenance cost, and *PSV* is the present value of scrap

[56]. Detailed values of each item of Eq. (32) are shown and discussed in the following sections.

A. Initial Capital Cost

The initial cost is collecting the price of all components of the wind energy system based on the market price for accurate results of the generated energy cost. The total cost of the WES is called the initial capital cost (IC). The IC is the sum of the price of WTs, installation cost, price of other components like a transmission line, transformers, and protection system. The civil work cost, and the components used in the wind energy system are estimated to be 20% of the price of WTs [57]. Based on these assumptions, the initial cost is determined from (33).

$$IC = 1.2 * WT_P * P_R * NWT + P_{inv} * INV_P \tag{33}$$

where WT_P is the price of kW of the rated power of WT, P_r is the rated power of the WT, NWT is the total number of WT required for the wind energy system. P_{inv} is the rated power of the inverter, and INV_P is the price of inverter per kW.

B. Replacement Cost

There are many components of the wind energy system that should be replaced during the lifetime of the project. These components should be determined at the start of the project and its present value at the beginning of the project should be determined. The net present values of components (Replacement cost, RC) should be calculated based on the interest rate (r) and inflation rate (i) as shown in (34) [57, 58]:

$$RC = \sum_{j=1}^{N_{rep}} \left(C_{RC} * C_U * \left(\frac{1+i}{1+r} \right)^{T*j/(N_{rep}+1)} \right) \tag{34}$$

where, C_{RC} is the capacity of the replacement unit (kW for WTs, and inverters, kWh, etc.), C_U is the unit price cost ($/kW for WT and inverters, $/kWh for battery), N_{rep} is the number of replacements during the system life period T.

C. Operation and Maintenance Cost

Many components of the wind energy system need an effective maintenance program during the lifetime of the project. The cost of maintenance during the lifetime of the project should be estimated and its net present value should be determined. Two parts of the wind energy systems need intensive maintenance programs which are the WTs and the power electronics converter. The maintenance of the WTs is estimated as 5% of the total cost of WTs, meanwhile, the maintenance cost of the inverter is estimated to be 1% [59]. Another study estimated the operation and maintenance cost (OMC) of the whole wind energy system to be 1% of the total cost of the wind energy system [56]. Some other studies [60] used fixed cost for each kW of the components such as

100 \$/kW for WTs and 1% of the cost of the inverter. The study shown in [61] used the annual maintenance cost of WT of \$20/kW. Reference [62] used the annual MOC to be \$10/kW for WTs. Some studies linked the OMC to the operation of the wind energy system by estimating the OMC as 1% of the total cost of kWh generated from the wind energy systems [63, 64]. The present value of OMC of the wind energy system used the total amount of maintenance and it uses the interest rate and inflation rate to estimate its value as shown in (36) [54]:

$$OMC = OMC_0 * \left(\frac{1+i}{r-i}\right) * \left(1 - \left(\frac{1+i}{1+r}\right)^T\right), \quad r \neq i \tag{35}$$

D. The Present Salvage Value

The replacements during the operation of the project and the price of the components at the end of the project should be taken into considerations. The price of these components should be determined as revenue at the start of the project taking into consideration the interest rate and inflation rate as shown in (36) [14, 56]. This salvage value (SV) is estimated to be 10% of WT, 20% of the power electronics converters and batteries, and other components will not have a value at the end of its use [14]. In this study, the SV has been taken as 20% for WTs, and 10% for the power conditioning components. The present value of SV should be calculated at the beginning of the project taking into considerations the interest rate and failure rate which is called the present salvage value (PSV) as shown in (36)

$$PSV = \sum_{j=1}^{N_{rep}+1} SV * \left(\frac{1+i}{1+r}\right)^{T*j/(N_{rep}+1)} \tag{36}$$

where SV is the scrap value, i is the inflation rate, r interest rate, N_{rep} is the number of different components replacements over the system life period T.

3.6 Selecting the Best Site and Best Wind Turbine for each Site "Fifth Subroutine"

This part of the program is used to extract the cost results of kWh generated from all options of sites and WTs as shown in Fig. 8. The minimum cost will be selected as the best option and the site and WT associate with this lowest cost will be selected to be used as the best matching between the sites and WT types. Many other useful results can be extracted from this subroutine like the best WT for each site based on the minimum cost for each site.

4 Output Results

The new proposed program has been used with 32 sites in Saudi Arabia and 140 market available WTs. The program can perform perfect site matching with WT type. The selection of suitable WT type has been carried out depending on the minimum energy price. An accurate cost estimation technique has been introduced to get accurate results. Many information can be extracted from this computer program. Some salient results have been presented in this report. Enormous helpful results can be also extracted from this program and it will not be displayed in this report due to the limitation of the report size. The enormous input data (32 sites wind speed data and 140 WT performance parameters) can perform an accurate wind energy map for Saudi Arabia. An excel program is used to implement the logic of this program to validate the results obtained from this subroutine. The output results from this computer program are shown in the following:

The 32 sites and the 140 WT types are used to draw a relationship between the rated wind speed and cut-in wind speeds as shown in Fig. 10. Using curve fitting of this relation to the first-order polynomial, the relation shown in (37) can be used to model this relation. The rated wind speed is directly proportional to the cut-in wind speed and the minimum and maximum cut-in wind speed are 2 m/s and 4 m/s, respectively. Meanwhile, most of the cut-in wind speed of the WTs is 3 m/s. Moreover, the minimum and maximum values of rated wind speed of WTs are 10 m/s and 15 m/s, respectively.

$$u_r = 1.2679 \, u_c + 8.5227 \tag{37}$$

The relationship between the Weibull scale parameter, c against the hub heights of WTs in five sites is shown in Fig. 11. It is clear from this figure that the Weibull scale parameter is linearly proportional to the hub height of the WTs.

Fig. 10 The rated wind speed along with the cut-in wind speed of WTs

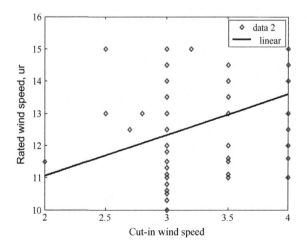

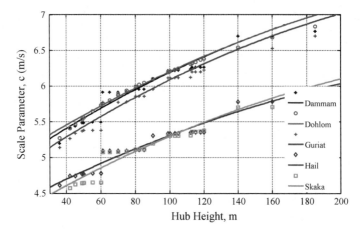

Fig. 11 The variation of Weibull scale parameter, c, and the hub height of WT, *h* for the best five sites among the 32 sites under study

Another very useful relationship between *c* and *ur* can be obtained from Fig. 12. This relation shows that the relationship showing *c* and u_r is linearly proportional. This relation is shown in Eq. (38). This relation is very helpful to the researchers and designers to determine the Weibull scale parameter directly from average wind speed without using the sophisticated calculations shown in the first subroutine.

$$c = 1.11 * u_{av} - 0.58 \tag{38}$$

Similar studies have been introduced in the literature to get this important relationship between the Weibull scale factor and the average wind speed of the site. One of these relations used the Gamma function to get another relation between average wind speed and Weibull parameters as shown in Eq. (39) [2]. Another equation introduced in (40) to model this relation between the Weibull scale parameters

Fig. 12 The variation of the Weibull scale parameter, *c* along with average wind speed for WTs and sites under study

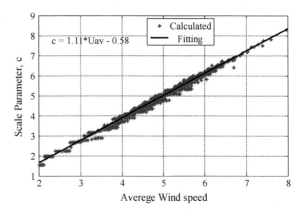

and average wind speed [65].

$$\frac{u_{av}}{c} = \Gamma(1 + \frac{1}{k}) \tag{39}$$

$$c = 1.12 * u_{av} \tag{40}$$

The Weibull parameters are used to determine the capacity factor of the wind energy system. The value of the capacity factor is very important where it is counted as the technical matching factor because the high value of the capacity factor means the high performance of the WT in this site. So, the capacity factor can be drawn based on the average wind speed and the rated wind speed as shown in Fig. 13. This relation is extracted from using the 32 sites and the 140 WTs used in this study. This relation shows that the highest capacity factor can be achieved at high average wind speed and low rated wind speed of the WT and vice versa.

The relation between the capacity factor, C_F and u_{av}/u_r for all sites under study and all WT types is shown in Fig. 14. This figure implies that the capacity factor increases with increasing the value of u_{av}/u_r for all sites and all WT types. So it is recommended to use sites with the highest average wind speed, and it is also recommended to use WT with the lowest rated wind speed to get the highest capacity factor. The highest capacity factor can be translated into lower cost and lower price of the generated kWh. It is also clear from this figure that the capacity factor increases with increasing the ratio of average and rated wind speed till u_{av}/u_r equal approximately to 1.22 after that the capacity factor will be reduced. So it is not recommended to install WT in any site with an average wind speed greater than 1.22 times the average wind speed.

Figure 15 shows the relationship between the energy price and capacity factor for 32 sites and 140 WT types. The energy cost is inversely proportional to the capacity

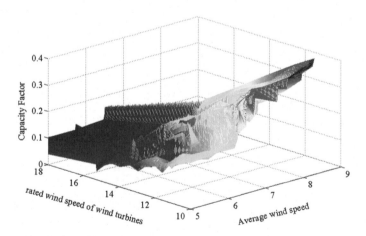

Fig. 13 3-D graph showing the relation between the capacity factor along with the rated wind speed of the WT and the average wind speed of the sites under study

Fig. 14 The relation between u_{av}/u_r and capacity factor

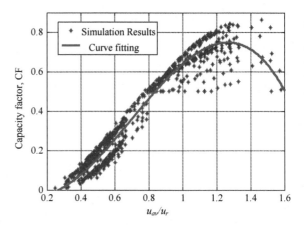

Fig. 15 The energy cost for the 32 sites and 140 WTs under study along with the capacity factor

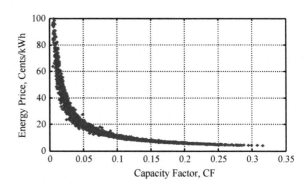

factor i.e., as the capacity factor increases the energy price decreases. Also, Fig. 16 shows the relation between the energy price (Cents/kWh) and the (u_{av}/u_r) for 32 sites and 140 WT types. This figure emphasizes that the energy price is inversely proportional to (u_{av}/u_r).

Fig. 16 The energy price in Cents/kWh for 32 sites and 140 WT along with (u_{av}/u_r)

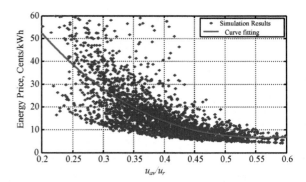

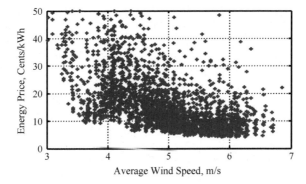

Fig. 17 The relation between the energy price and average wind speed for 32 sites and 140 WTs

Figure 17 shows the relation between the energy price and average wind speed for 32 sites and 140 WT types. The energy price is inversely proportional to the average wind speed i.e., as the average wind speed increases the energy price decreases. As an example, Fig. 18 shows the relation between energy price (Cents/kWh) and the average wind speed for 32 sites with WT #1. This figure emphasizes this inverse relationship between energy price and average wind speed.

Table 3 shows site information in order from best to worst depending on *ECF*. It is clear from this table that the best site among these 32 sites is Dammam with 4.155073 Cents/kWh and the next four sites are Douhlom, Guriat, Hail, Skaka with 4.395947, 4.395947, 4.576602, and 5.239005 Cents/kWh, respectively. These best five sites results will be discussed in detail and will be compared with other sites in the following discussion. The most interesting points in these results are that all of these sites do not have the highest average wind speed but at least these sites have good wind speed with respect to other sites. Also, the price of kWh generated from Douhlom and Guriat sites with WT #2 is 5.8% greater than the price of kWh

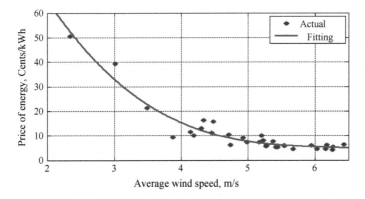

Fig. 18 The relation between energy price Cents/kWh and the average wind speed for 32 sites with WT #1 as an example

Table 3 Sites information in order from best to worst depending on ECF

Site Order	Site name	Best WT		Average Wind speed	ECF	% increase from the best site,
		WT #	WT Name	m/s	C/kWh	'Dammam'
1	Dammam	2	Goldwind_3	4.4	4.155073	0.0
2	Douhlom	2	Goldwind_3	4.32	4.395947	5.8
3	Guriat	2	Goldwind_3	4.24	4.395947	5.8
4	Hail	1	Acciona_6	3.98	4.576602	10.1
5	Skaka	2	Goldwind_3	3.82	5.239005	26.1
6	Asulil	1	Acciona_6	3.8	5.299224	27.5
7	KFIA	1	Acciona_6	4.41	5.419661	30.4
8	Al-Ahsa	1	Acciona_6	3.7	5.660534	36.2
9	Al-Jouf	1	Acciona_6	3.9	5.780971	39.1
10	Dhahran	3	Goldwind_4	4.31	6.021845	44.9
11	Rafha	1	Acciona_6	3.89	6.021845	44.9
12	Turaif	1	Acciona_6	4.17	6.021845	44.9
13	Qaisumah	1	Acciona_6	3.71	6.142282	47.8
14	Hafer-Albate	1	Acciona_6	3.33	6.142282	47.8
15	Wejh	1	Acciona_6	4.52	6.383155	53.6
16	Wadi	1	Acciona_6	3.5	7.226214	73.9
17	Arar	1	Acciona_6	3.63	7.226214	73.9
18	Yanbou	1	Acciona_6	3.77	7.587525	82.6
19	Taif	1	Acciona_6	3.67	8.069273	94.2
20	Al-Baha	1	Acciona_6	3.46	9.032767	117.4
21	Tabouk	1	Acciona_6	2.72	9.273642	123.2
22	Riyadh	1	Acciona_6	2.95	9.996264	140.6
23	Jeddah	1	Acciona_6	3.65	9.996264	140.6
24	Sharorah	1	Acciona_6	3.31	10.47801	152.2
25	Abha	1	Acciona_6	3.13	11.20063	169.6
26	Gassim	1	Acciona_6	2.91	11.44151	175.4
27	Khamis	1	Acciona_6	3.02	13.00718	213.0
28	Madinah	1	Acciona_6	3.05	16.25898	291.3
29	Bisha	1	Acciona_6	2.45	21.1969	410.1
30	Gizan	63	Fuhrlander_4	3.15	29.00522	598.1
31	Najran	1	Acciona_6	2.12	39.38287	847.8
32	Makkah	34	AWE_1	1.64	129.8009	3023.9

generated from the Dammam site with WT #2. Also, it is clear that the price of kWh generated from the Hail site with WT #1 is 10.1% greater than the price of kWh generated from the Dammam site with WT #2. Also, the price of kWh generated from the Skaka site with WT #2 is 26.1% greater than the price of kWh generated from the Dammam site with WT #2. Also, it is clear from the table that the worst sites in wind energy applications are Madinah, Bisha, Gizan, Najran, and Makkah where the price of kWh generated from these sites with the best WT for each of them are 16.25898, 21.1969, 29.00522, 39.38287, and 129.8009 Cents/kWh, respectively. These sites are not good option to install the WES on it. Also, it is noted that the cost of the generated kWh from the WES in these five sites are almost 4, 5, 7, 9.5, 31 times the cost of kWh generated from the Dammam site with WT #2.

Many interesting information and recommendations can be concluded from Table 4. The following points summarized the salient information and recommendations that can be extracted from the above table.

Table 4 Sites in order from best to worst depending on ECF

Site Order	Site name	Location	Average Speed	1	2	3	4	5	6	7	8	9	10
								Best WT					
1	Dammam	EM	4.4	2	1	113	3	54	53	17	88	99	19
2	Dohlom	WS	4.32	2	1	3	113	54	53	97	17	99	63
3	Guriat	WN	4.24	2	1	3	113	54	53	17	19	88	97
4	Hail	MN	3.98	1	2	113	3	54	53	88	17	99	63
5	Skaka	MN	3.82	2	1	113	3	54	53	17	88	97	99
6	Asulil	MM	3.8	1	2	113	3	54	53	88	17	97	99
7	KFIA	WN	4.41	1	3	2	54	113	53	63	17	99	97
8	Al-Ahsa	EM	3.7	1	2	113	3	54	53	88	17	99	19
9	Al-Jouf	MN	3.9	1	2	113	3	54	53	88	17	19	97
10	Dhahran	EM	4.31	3	1	2	54	113	17	97	99	113	63
11	Rafha	WN	3.89	1	2	3	113	54	53	63	19	97	17
12	Turaif	WN	4.17	1	2	3	54	113	53	17	19	97	99
13	Qaisumah	EN	3.71	1	2	113	3	54	53	88	19	17	99
14	Hafer Al-Baten	EN	3.33	1	2	113	3	35	36	34	54	53	88
15	Wejh	WN	4.52	1	3	54	2	53	17	99	19	97	113
16	Wadi	MS	3.5	1	2	113	3	54	53	88	19	17	99
17	Arar	WN	3.63	1	2	113	3	54	53	88	63	19	97
18	Yanbou	WM	3.77	1	2	3	113	54	53	63	19	97	17
19	Taif	WM	3.67	1	2	113	3	63	54	53	19	17	88
20	Al-Baha	WN	3.46	1	2	113	3	54	88	63	19	53	17
21	Tabouk	WN	2.72	1	2	35	36	34	113	3	54	88	53
22	Riyadh	MM	2.95	1	2	113	3	54	88	53	34	35	36
23	Jeddah	WM	3.65	1	2	3	113	63	54	19	53	88	17
24	Sharorah	ES	3.31	1	2	113	3	54	88	63	19	53	17
25	Abha	WS	3.13	1	2	113	63	3	88	54	19	53	29
26	Gassim	MM	2.91	1	2	113	3	88	54	53	34	35	36
27	Khamis	WS	3.02	1	2	113	3	88	63	54	19	53	29
28	Madinah	WM	3.05	1	2	113	63	88	3	19	54	29	53
29	Bisha	MM	2.45	1	2	113	88	35	36	34	3	86	54
30	Gizan	WS	3.15	63	2	1	113	88	67	19	3	29	54
31	Najran	WS	2.12	1	2	34	35	36	113	88	3	86	54
32	Makkah	WM	1.64	34	35	36	1	2	113	122	3	86	67

Hint: E=East, W=West, M=Middle, N=North, S=South

- WTs # 1, 2, 3, 113, and 54 are shown among the best 10 WT types for all sites under study. So, it is recommended to use any one of these WT types for sites that do not have a design study or unavailable wind speed data.
- The best two WTs in most sites are WTs # 1 and 2. This is the case in 30 sites except for Gizan and Makkah because these two sites have very low wind speeds. So, it is recommended to use WTs # 1, 2 in most of the sites in Saudi Arabia except sites with very low average wind speed.
- WT # 2 is the best selection for sites with good average wind speeds such as Dammam, Douhlom, Guriat, and Skaka. Also, the second option for these sites is WT #1. So, it is recommended to use WT #2 in sites with average wind speed greater than 4 m/s in 10 m elevation.
- WT #1 is the best option for sites with medium average wind speed and WT #2 is the second-best option in most of these cases. So, it is recommended to use WT #1 in sites with average wind speed lower than 4 m/s and greater than 2.45 m/s in 10 m' elevation except for Gizan, Dhahran, and Skaka.
- WTs # 34, 35, and 36 are the best options for the sites having very low average wind speeds such as Makkah and Najran. Where these WT types were never shown among the best 10 WTs except with sites having average wind speed lower than 3 m/s. So, it is recommended to use WTs # 34, 35, and 36 for sites lower

than 2.4 m/s and it is acceptable to use them for average wind speed between 2.4 and 3 m/s over 10 m elevation.

4.1 Detailed Results from Best Five Sites

Detailed results of the best five sites are shown in the following sections. These results are focused on the comparison between the best and worst WT types; also a comparison between the five best and five worst WTs for each site will be displayed. Also, a list for the best 10 WT types and 10 worst WT types will be shown in tables for these five sites.

4.1.1 Detailed Results from Dammam Site

It is clear from the results of the new proposed computer program that has been summarized in Tables 5 and 6 that the Dammam site is the best site for wind energy applications. The best WT type for the Dammam site is WT #2. The price of kWh generated from Dammam with WT #2 is 4.155073 Cents/kWh. The worst WT that can be used in Dammam is NEPC_2 (WT #125) with 12.766 Cents/kWh which shows that it is three times the price of kWh generated in Dammam with WT #2. The main reason in the big difference in the price of kWh is because most of the available wind speed of the Dammam site is used to generate power from WT #2 but most of these

Table 5 The sest 10 WT types in the Dammam site

WT		u_c	u_r	u_f	c	k	C_F	ECF
Name	#							
Goldwind_3	2	3	10.3	22	6.220	1.709	0.315	4.155
Acciona_6	1	3	10.6	20	6.264	1.695	0.307	4.215
HZ_WindPR_3	113	3	10.5	25	6.220	1.709	0.307	4.255
Goldwind_4	3	3	9.9	22	5.918	1.790	0.299	4.276
Envision_4	54	3	10	25	5.913	1.792	0.294	4.396
Envision_3	53	3	10.5	25	5.957	1.772	0.279	4.456
Acciona_3	17	3	10.5	20	5.956	1.772	0.278	4.516
Goldwind_2	88	3	11	22	6.220	1.709	0.288	4.516
Guodian_3	99	3	10.5	25	5.956	1.772	0.279	4.516
Acciona_5	19	3.5	11	25	6.264	1.695	0.276	4.577

Table 6 The detailed results for the worst 10 WT types in the Dammam site

WT Name		u_c	u_r	u_f	c	k	C_F	ECF
Name	#							
GC_China_Turbine_Corp_1	74	4	15	25	5.913	1.792	0.126	9.153
GC_China_Turbine_Corp_2	75	4	15	25	5.913	1.792	0.126	9.153
GC_China_Turbine_Corp_3	76	4	15	21	5.913	1.792	0.126	9.193
Gamesa_2	68	4	15	25	5.912	1.792	0.126	9.380
Gamesa_3	69	4	15	25	5.912	1.792	0.126	9.384
Ecotecnia_2	44	4	14.5	25	5.440	1.681	0.119	9.659
NEPC_1	124	4	15	25	5.440	1.681	0.111	10.234
Southern_Wind_Farms	128	4	15	25	5.440	1.681	0.111	10.234
NEPC_3	126	4	15	25	5.198	1.746	0.092	11.305
NEPC_2	125	4	17	25	5.440	1.681	0.089	12.766

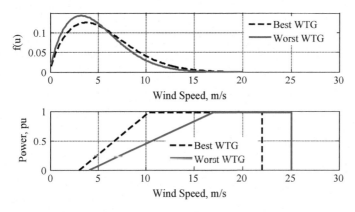

Fig. 19 Comparison between the best and worst WT types for the Dammam site

speeds are not used to generate power from WT #125. These conclusions are shown very clear in Fig. 19. This figure shows a comparison between the best WT type for Dammam (WT #2) and the worst WT type for this site (WT #125). The first curve shows the speed distribution facing each WT type in the Dammam site. The second curve shows the generated power against wind speed for WT #2 and WT #125. It is clear from this figure that most of the speed distribution for Dammam is lying within the generated area of WT #2. But, most of the speeds are shown below the cut-in speed of WT #125 and there is no considerable speed over the rated speed of WT #125. The same results can be concluded from the next two figures where Fig. 20 and Fig. 21 show the distribution of wind speed and the power characteristics against wind speed for the best and worst five WTs for the Dammam site, respectively.

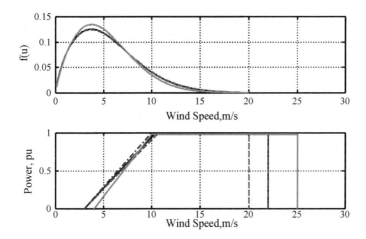

Fig. 20 Best five WT types in the Dammam site

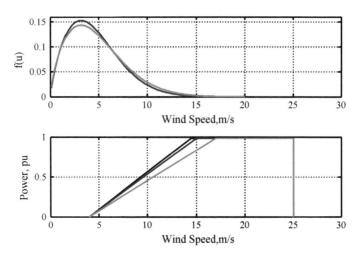

Fig. 21 Worst five WT types for Dammam site

These two figures show that most of the speed in the Dammam site is lying within the generated power of five best WTs but most of the speeds are not lying within the generated power area of the worst five WTs for the Dammam site.

Tables 5 and 6 show the list of the best 10 WTs and the worst 10 WTs for Dammam. These tables also show the WTs name, numbers, specifications, Weibull parameters, capacity factor, and the cost of energy generated from each WT. Moreover, the cost of energy is inversely proportional to the capacity factor in each case. Also, it is better to use WT #2 in the Dammam site for the minimum cost of generated energy. Also if a wrong WT type like WT #125 is used instead of WT #2 in Dammam site, the price of the generated kWh will be more than three times the price associated with WT #2. So it is not recommended to use any one of the worst WT types shown in Table 6 on the Dammam site.

4.1.2 Detailed Results from Douhlom Site

It is clear from the results of the new proposed computer program that has been summarized in Tables 7 and 8 that the Douhlom site is the second best site for wind energy applications. The best WT type for the Douhlom site is WT #2. The price of kWh generated from Douhlom with WT #2 is 4.396 Cents/kWh. The worst WT that can be used in Douhlom is NEPC_2 (WT #125) with 19.36 Cents/kWh which is clear that it is more than four times the price of kWh generated in Douhlom with WT #2. The main reason of the big difference in the price of kWh is because most of the available wind speed of the Douhlom site is used to generate power from WT #2 but most of these speeds are not used to generate power from WT #125. These conclusions are shown very clear in Fig. 22. This figure shows a comparison between the best WT type for Douhlom (WT #2) and the worst WT type for this

Table 7 The detailed results for the best 10 WT types in the Douhlom site

WT		u_c	u_r	u_f	c	k	C_F	ECF
Name	#							
Goldwind_3	2	3	10.3	22	6.206	2.130	0.277	4.396
Acciona_6	1	3	10.6	20	6.371	2.137	0.277	4.456
Goldwind_4	3	3	9.9	22	5.942	2.120	0.272	4.456
HZ_WindPR_3	113	3	10.5	25	6.206	2.130	0.267	4.577
Envision_4	54	3	10	25	5.872	2.115	0.260	4.637
Envision_3	53	3	10.5	25	6.052	2.127	0.253	4.757
Guodian_1	97	3	10.5	25	5.999	2.127	0.248	4.817
Acciona_3	17	3	10.5	20	5.999	2.127	0.248	4.878
Guodian_3	99	3	10.5	25	5.999	2.127	0.248	4.878
Fuhrlander_4	63	3.5	11.5	25	6.685	2.150	0.250	4.918

Table 8 The detailed results for the worst 10 WT types in the Douhlom site

WT		u_c	u_r	u_f	c	k	C_F	ECF
Name	#							
GC_China_Turbine_Corp_2	75	4	15	25	5.872	2.115	0.094	12.285
GC_China_Turbine_Corp_3	76	4	15	21	5.872	2.115	0.094	12.285
Ecotecnia_3	45	4	14.5	25	5.656	2.110	0.091	12.586
Gamesa_3	69	4	15	25	5.805	2.111	0.091	12.660
Gamesa_2	68	4	15	25	5.805	2.111	0.091	12.665
Ecotecnia_2	44	4	14.5	25	5.472	2.097	0.083	13.633
NEPC_1	124	4	15	25	5.472	2.097	0.077	14.696
Southern_Wind_Farms	128	4	15	25	5.472	2.097	0.077	14.696
NEPC_3	126	4	15	25	5.279	2.083	0.069	16.090
NEPC_2	125	4	17	25	5.472	2.097	0.058	19.360

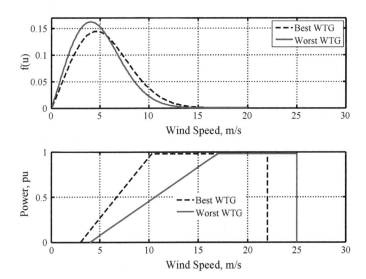

Fig. 22 Comparison between the best and worst WT types for the Douhlom site

site (WT #125). The first curve shows the speed distribution facing each WT in the Douhlom site. The second curve shows the generated power against wind speed for WT #2 and WT #125. It is clear from this figure that most of the speed distribution for Douhlom is lying within the generated area of WT #2. But, most of the speeds are shown below the cut-in speed of WT #125 and there is no considerable speed over the rated speed of WT #125. The same results can be concluded from the next two figures, Fig. 23 and Fig. 24, which show the distribution of wind speed and the power characteristics against wind speed for the best and worst five WT types in the Douhlom site, respectively. These two figures show that most of the speed in the Douhlom site is lying within the generated power of the five best WT types but most of the speeds are not lying within the generated power area of the worst five WT types in Douhlom site.

Tables 7 and 8 show the list of the best 10 WTs and the worst 10 WTs for the Douhlom site. These tables also show the WT names, numbers, specifications, Weibull parameters, capacity factor, and cost of generated energy with each WT type.

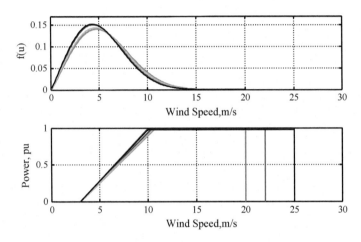

Fig. 23 Best five WT types in the Douhlom site

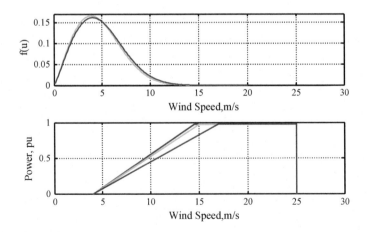

Fig. 24 Worst five WT types for Douhlom

Moreover, the prices of energy generated are inversely proportional to the capacity factor in each case. Also, it is better to use WT #2 in the Douhlom site for the minimum cost of generated energy. Also if a wrong WT type like WT #125 is used instead of WT #2 in Douhlom, the price of the generated kWh will be more than four times the price associated with WT #2. So it is not recommended to use any one of the worst WT types shown in Table 8 in the Douhlom site.

4.1.3 Detailed Results from Guriat Site

It is clear from the results of the new proposed computer program that has been summarized in Table 9 and Table 10 that the Guriat site is the third best site for wind energy applications. The best WT type for the Guriat site is WT #2. The price of kWh generated from Guriat with WT #2 is 4.396 Cents/kWh. The worst WT that can be used in Guriat is NEPC_2 (WT #125) with 18.708 Cents/kWh which is clear that it is four times the price of kWh generated in Guriat with WT #2. The main reason of the big difference in the price of kWh is because most of the available wind speed of the Guriat site is used to generate power from WT #2 but most of these speeds are not used to generate power from WT #125. These conclusions are shown very clear in Fig. 25. This figure shows a comparison between the best WT for Guriat (WT #2) and the worst WT for this site (WT #125). The first curve shows the speed distribution facing each WT in the Guriat site. The second curve shows the generated power against wind speed for WT #2 and WT #125. It is clear from

Table 9 The detailed results for the best 10 WT types for the Guriat site

WT		u_c	u_r	u_f	c	k	C_F	ECF
Name	#							
Goldwind_3	2	3	10.3	22	6.121	1.959	0.284	4.396
Acciona_6	1	3	10.6	20	6.201	1.931	0.281	4.456
Goldwind_4	3	3	9.9	22	5.782	2.066	0.261	4.577
HZ_WindPR_3	113	3	10.5	25	6.121	1.959	0.275	4.577
Envision_4	54	3	10	25	5.779	2.067	0.256	4.757
Envision_3	53	3	10.5	25	5.864	2.016	0.246	4.817
Acciona_3	17	3	10.5	20	5.864	2.016	0.246	4.938
Acciona_5	19	3.5	11	25	6.201	1.931	0.249	4.938
Goldwind_2	88	3	11	22	6.121	1.959	0.254	4.938
Guodian_1	97	3	10.5	25	5.864	2.016	0.246	4.938

Table 10 The detailed results for the worst 10 WT types in Guriat site

WT		u_c	u_r	u_f	c	k	C_F	ECF
Name	#							
GC_China_Turbine_Corp_1	74	4	15	25	5.779	2.067	0.093	12.686
GC_China_Turbine_Corp_2	75	4	15	25	5.779	2.067	0.093	12.686
GC_China_Turbine_Corp_3	76	4	15	21	5.779	2.067	0.093	12.686
Gamesa_2	68	4	15	25	5.779	2.067	0.093	13.089
Gamesa_3	69	4	15	25	5.779	2.067	0.093	13.104
Ecotecnia_2	44	4	14.5	25	5.343	1.948	0.088	13.441
NEPC_1	124	4	15	25	5.343	1.948	0.082	14.416
Southern_Wind_Farms	128	4	15	25	5.343	1.948	0.082	14.416
NEPC_3	126	4	15	25	5.145	2.012	0.068	16.379
NEPC_2	125	4	17	25	5.343	1.948	0.063	18.708

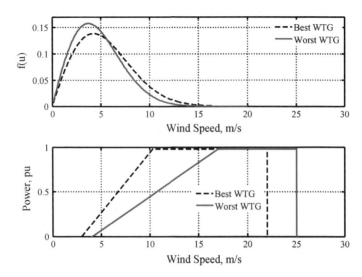

Fig. 25 Comparison between the best and worst WT types for the Guriat site

this figure that most of the speed distribution for Guriat is lying within the generated area of WT #2. But, most of the speeds are shown below the cut-in speed of WT #125 and there is no considerable speed over the rated speed of WT #125. The same results can be concluded from the next two figures where Fig. 26 and Fig. 27 show the distribution of wind speed and the power characteristics against wind speed for the best and worst five WTs in the Guriat site, respectively. These two figures show that most of the speeds in the Guriat site are lying within the generated power of five

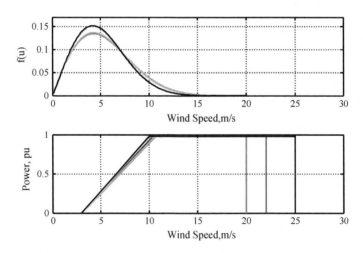

Fig. 26 Best five WT types in the Guriat site

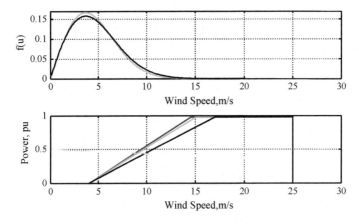

Fig. 27 Worst five WT types for the Guriat site

best WT types but most of the speeds are not lying within the generated power area of the worst five WT types in the Guriat site.

Tables 9 and 10 show the list of the best 10 WTs and the worst 10 WTs for the Guriat site. These tables also show the WT names, numbers, specifications, Weibull parameters, capacity factor, and the price of kWh generated with each WT type. It is also clear from these two tables that the prices of kWh generated are inversely proportional to the capacity factor in each case. Also, it is recommended to use WT #2 in the Guriat site for the minimum price of kWh generated. Also if a wrong WT type like WT #125 is used instead of WT #2 in Guriat the price of the generated kWh will be almost four times the price associated with WT #2. So it is not recommended to use any one of the worst WTs shown in Table 10 in the Guriat site.

4.1.4 Detailed Results from Hail Site

It is clear from the results of the new proposed computer program that has been summarized in Tables 11 and 12 that the Hail site is the fourth best site for wind energy applications. The best WT type for the Hail site is WT #1. The price of kWh

Table 11 The detailed results for the best 10 WT types in the Hail site

WT		u_c	u_r	u_f	c	k	C_F	ECF
Name	#							
Acciona_6	1	3	10.6	20	5.357	1.429	0.258	4.577
Goldwind_3	2	3	10.3	22	5.338	1.434	0.267	4.577
HZ_WindPR_3	113	3	10.5	25	5.338	1.434	0.261	4.657
Goldwind_4	3	3	9.9	22	5.096	1.518	0.251	4.697
Envision_4	54	3	10	25	5.094	1.519	0.247	4.878
Envision_3	53	3	10.5	25	5.111	1.515	0.233	4.938
Goldwind_2	88	3	11	22	5.338	1.434	0.245	4.938
Acciona_3	17	3	10.5	20	5.109	1.515	0.232	4.998
Guodian_3	99	3	10.5	25	5.109	1.515	0.233	4.998
Fuhrlander_4	63	3.5	11.5	25	5.782	1.497	0.242	5.018

Table 12 The detailed results for the worst 10 WT types in the Hail site

WT		u_c	u_r	u_f	c	k	C_F	ECF
Name	#							
GC_China_Turbine_Corp_3	76	4	15	21	5.094	1.519	0.111	10.157
Gamesa_3	69	4	15	25	5.093	1.519	0.111	10.408
Gamesa_2	68	4	15	25	5.093	1.519	0.111	10.413
W.T.S._4,	7	7.1	16.2	27	6.415	1.379	0.118	10.598
Ecotecnia_3	45	4	14.5	25	4.778	1.566	0.094	10.598
Ecotecnia_2	44	4	14.5	25	4.743	1.577	0.091	11.393
NEPC_1	124	4	15	25	4.743	1.577	0.086	12.059
Southern_Wind_Farms	128	4	15	25	4.743	1.577	0.086	12.059
NEPC_3	126	4	15	25	4.614	1.634	0.074	13.312
NEPC_2	125	4	17	25	4.743	1.577	0.069	14.914

generated from Hail with WT #1 is 4.577 Cents/kWh. The worst WT type that can be used in Hail is NEPC_2 (WT #125) with 14.914 Cents/kWh which is clear that it is more than three times the price of kWh generated in Hail with WT #1. The main reason of the big difference in the price of kWh is because most of the available wind speed of the Hail site is used to generate power from WT #1 but most of these speeds are not used to generate power from WT #125. These conclusions are shown very clear in Fig. 28. This figure shows a comparison between the best WT type for Hail (WT #1) and the worst WT type for this site (WT #125).

The first curve shows the speed distribution facing each WT type in the Hail site. The second curve shows the generated power against wind speed for WT #1 and WT #125. It is clear from this figure that most of the speed distribution for Hail is lying within the generated area of WT #1. But most of the speeds are shown below the cut-in wind speed of WT #125 and there is no considerable speed over the rated speed of WT #125. The same results can be concluded from the next two figures where Fig. 29 and Fig. 30 show the distribution of wind speed and the power characteristics

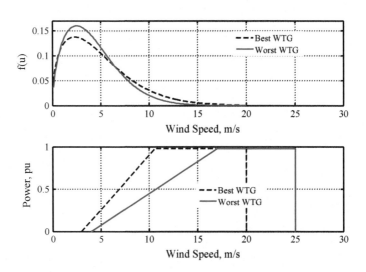

Fig. 28 Comparison between the best and worst WT types for the Hail site

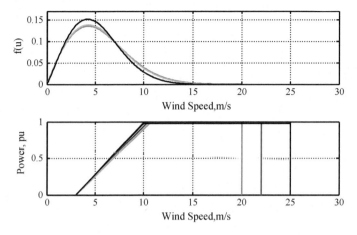

Fig. 29 Best five WT types in the Hail site

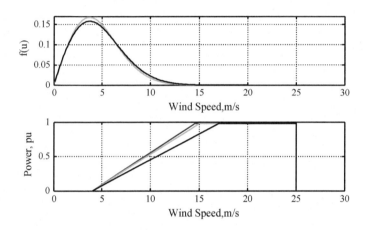

Fig. 30 Worst five WT types in the Hail site

against wind speed for the best and worst five WT types in the Hail site respectively. These two figures show that most of the speed in the Hail site is lying within the generated power of five best WTs but most of the speeds are not lying within the generated power area of the worst five WTs in the Hail site.

Tables 11 and 12 show the list of the best 10 WT types and the worst 10 WT types for the Hail site. These tables also show the WT names, numbers, specifications, Weibull parameters, capacity factor, and the cost of generated energy from each WT type. Moreover, the cost of generated energy is inversely proportional to the capacity factor in each case. Also, it is better to use WT #1 in the Hail site for the minimum price of kWh generated. Also if a wrong WT type like WT #125 is used instead of WT #1 in Hail the cost generated energy will be almost three times the cost associated

Table 13 The detailed results for the best 10 WT types in the Skaka site

WT		u_c	u_r	u_f	c	k	C_F	ECF
Name	#							
Goldwind_3	2	3	10.3	22	5.309	1.690	0.237	5.239
Acciona_6	1	3	10.6	20	5.369	1.662	0.235	5.299
HZ_WindPR_3	113	3	10.5	25	5.309	1.690	0.230	5.380
Goldwind_4	3	3	9.9	22	5.086	1.768	0.223	5.420
Envision_4	54	3	10	25	5.081	1.769	0.219	5.600
Envision_3	53	3	10.5	25	5.116	1.754	0.206	5.721
Acciona_3	17	3	10.5	20	5.114	1.754	0.206	5.781
Goldwind_2	88	3	11	22	5.309	1.690	0.214	5.781
Guodian_1	97	3	10.5	25	5.114	1.754	0.206	5.781
Guodian_3	99	3	10.5	25	5.114	1.754	0.206	5.781

Table 14 The detailed results for the worst 10 WT types in the Skaka site

WT		u_c	u_r	u_f	c	k	C_F	ECF
Name	#							
GC_China_Turbine_Corp_2	75	4	15	25	5.081	1.769	0.084	12.726
GC_China_Turbine_Corp_3	76	4	15	21	5.081	1.769	0.084	12.766
Ecotecnia_2	44	4	14.5	25	4.627	1.510	0.092	12.790
Gamesa_3	69	4	15	25	5.080	1.770	0.084	13.069
Gamesa_2	68	4	15	25	5.080	1.770	0.084	13.089
W.T.S._4,	7	7.1	16.2	27	6.465	1.752	0.079	13.328
NEPC_1	124	4	15	25	4.627	1.510	0.087	13.522
Southern_Wind_Farms	128	4	15	25	4.627	1.510	0.087	13.522
NEPC_3	126	4	15	25	4.428	1.581	0.071	15.127
NEPC_2	125	4	17	25	4.627	1.510	0.071	16.650

with WT #1. So it is not recommended to use any one of the worst WT types shown in Table 12 in the Hail site.

4.1.5 Detailed Results from Skaka Site

It is clear from the results of the new proposed computer program that has been summarized in Tables 13 and 14 that the Skaka site is the fifth best site for wind energy applications. The best WT type for the Skaka site is WT #2. The price of kWh generated from Skaka with WT #2 is 5.239 Cents/kWh. The worst WT type that can be used in Skaka is NEPC_2 (WT #125) with 16.65 Cents/kWh which is clear that it is three times the price of kWh generated in Skaka with WT #2. The main reason of the big difference in the price of kWh is because most of the available wind speed of the Skaka site is used to generate power from WT #2 but most of these speeds are not used to generate power from WT #125. These conclusions are shown very clear in Fig. 31. This figure shows a comparison between the best WT type for Skaka (WT #2) and the worst WT type for this site (WT #125). The first curve shows the speed distribution facing each WT type in the Skaka site. The second curve shows the generated power against wind speed for WT #2 and WT #125. It is clear from this figure that most of the speed distribution for Skaka is lying within the generated area of WT #2. But, most of the speeds are shown below the cut-in wind speed of WT #125 and there is no considerable speed over the rated speed of WT #125. The same results can be concluded from the next two figures where Fig. 32 and Fig. 33 show

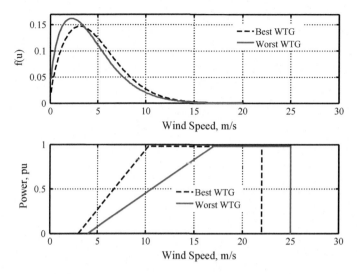

Fig. 31 Comparison between the best and worst WT types for the Skaka site

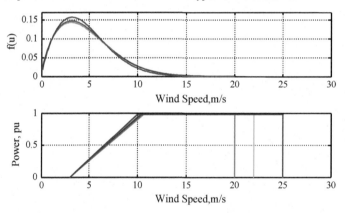

Fig. 32 Best five WT types in the Skaka site

the distribution of wind speed and the power characteristics against wind speed for the best and worst five WT types in the Skaka site, respectively. These two figures show that most of the speed in the Skaka site is lying within the generated power of the five best WTs but most of the speeds are not lying within the generated power area of the worst five WT types in the Skaka site.

Tables 13 and 14 show the list of the best 10 WT types and the worst 10 WT types for the Skaka site. These tables also show the WT names, numbers, specifications, Weibull parameters, capacity factor, and the cost of generated energy from each WT type. The prices of generated energy is inversely proportional to the capacity factor in each case. Also, it is better to use WT #2 in the Skaka site for the minimum cost of the generated energy. Also if a wrong WT type like WT #125 is used instead of WT #2 in Skaka the cost of the generated energy is almost three times the cost associated

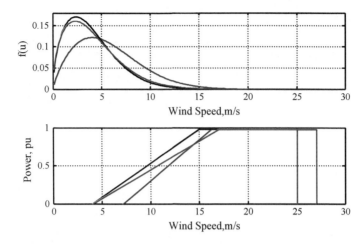

Fig. 33 Worst five WT types in the Skaka site

with WT #2. So it is not recommended to use any one of the worst WT types shown in Table 14 in the Skaka site.

5 Conclusions

The cost of energy generated from wind energy systems, WES, is depending on the sites that will be used to install the WES and the wind turbine (WT) used in this site. The WT can work with high performance in some sites but it may not work well in some other sites, and for this reason, it is essential to start selecting the best site among several sites available to install the WES and the suitable WT for this site. For this reason, the new proposed computer program is used to handle this optimization problem. 32 sites are selected in different provinces of Saudi Arabia and 140 market available WT types are selected to choose the best site and the best WT suitable for this site. The best site among these sites is the Dammam and the best WT suitable for this site is Goldwind_3. The cost of energy for this option is 4.16 Cents/kWh. Also, it is clear from the results of the proposed computer program the worst site among these sites understudy is the Makkah site where the cost of energy is 129.81 Cents/kWh when using the *AWE_1* WT. The big difference between the cost of energy improved the superiority to be used to choose the best site and the best WT.

References

1. Pallabazzer R (2004) Previsional estimation of the energy output of wind generators. Renew Energy 29:413–420
2. JohnsonGL (2003) Wind energy systems. Prentice-Hall Inc., England Cliffs
3. Eltamaly AM (2014) Pairing between sites and wind turbines for Saudi Arabia sites. Arab J Sci Eng 39(8):6225–6233
4. Hua S, Chengb J (2007) Performance evaluation of pairing between sites and wind turbines. Renew Energy 32:1934–1947
5. Rodman LC, Meentemeyer RK (2006) A geographic analysis of wind turbine placement in Northern California. Energy Policy J 34:2137–2149
6. Eltamaly AM, Mohamed MA (2014) A novel design and optimization software for autonomous PV/wind/battery hybrid power systems. Math Probl Eng
7. Eltamaly AM (2013) Design and implementation of wind energy system in Saudi Arabia. Renew Energy 60:42–52
8. Eltamaly AM, Addoweesh KE, Bawah U, Mohamed MA (2013) New software for hybrid renewable energy assessment for ten locations in Saudi Arabia. J Renew Sustain Energy 5(3):033126
9. Ahmed Shata AS, Hanitsch R (2006) The potential of electricity generation on the east coast of Red Sea in Egypt. Renew Energy 31:1597–1615
10. Eltamaly AM (2013) Design and simulation of wind energy system in Saudi Arabia. In: 2013 4th international conference on intelligent systems, modelling and simulation. IEEE, pp 376–383
11. Rehman S, Halawani T, Husain T (1994) Weibull parameters for wind speed distribution in Saudi Arabia. Solar Energy 53(6):473–479
12. Eltamaly AM, Farh HM (2012) Wind energy assessment for five locations in Saudi Arabia. J Renew Sustain Energy 4(2):022702
13. Ahmed Shata AS, Hanitsc R (2008) Electricity generation and wind potential assessment at Hurghada, Egypt. Renew Energy 33:141–148
14. Ahmed Shata AS, Hanitsc R (2006) Evaluation of wind energy potential and electricity generation on the coast of Mediterranean Sea in Egypt. Renew Energy 31:1183–1202
15. EL-Tamaly HH, Hamada M, EL-Tamaly AM (1995) Computer simulation of wind energy system and applications. In: Proceedings international AMSE conference in system analysis, control & design, vol 4, pp 84–93, Brno, Czech Republic, 3–5 July 1995
16. Bawah U, Addoweesh KE, Eltamaly AM (2012) Economic modeling of site-specific optimum wind turbine for electrification studies. Advanced materials research, vol 347. Trans Tech Publications Ltd., pp 1973–1986
17. Shamma'a A, Abdullrahman A, Addoweesh KE, Eltamaly A (2012) Optimum wind turbine site matching for three locations in Saudi Arabia. Advanced materials research, vol 347. Trans Tech Publications Ltd., pp 2130–2139
18. ENVIS Centre on Renewable Energy and Environment "Wind Energy Information". Technical report, 2005/2006
19. Ahmed MA, Eltamaly AM, Alotaibi MA, Alolah AI, Kim Y-C (2020) Wireless network architecture for cyber physical wind energy system. IEEE Access 8:40180–40197
20. Jungbluth N, Bauer C, Dones R, Frischknecht R (2004) Life cycle assessment for emerging technologies: case studies for photovoltaic and wind power. Int J Life Cycle Assess 10:24–34
21. Mohamed MA, Eltamaly AM (2018) Sizing and techno-economic analysis of stand-alone hybrid photovoltaic/wind/diesel/battery energy systems. Modeling and simulation of smart grid integrated with hybrid renewable energy systems. Springer, Cham, pp 23–38
22. Martinez E, Sanz F, Pellegrini S, Jimenez E, Blanco J (2009) Life-cycle assessment of a 2-MW rated power wind turbine: CML method. Int J Life Cycle Assess 14:52–63
23. Al-Saud MST, Eltamaly AMA, Al-Ahmari AMA (2017) Multi-rotor vertical axis wind turbine. U.S. Patent 9,752,556, 5 Sept 2017
24. Tande JO, Hunterv R (1994) Recommended practices for wind turbine testing: estimation of cost of energy from wind energy conversion system, 2nd edn

25. Manwell JF, McGowan JG, Rogers AL (2009) Wind energy explained—theory design and application, 2nd edn. ISBN 978-0-470-01500-1
26. Mohamed MA, Eltamaly AM, Alolah AI (2017) Swarm intelligence-based optimization of grid-dependent hybrid renewable energy systems. Renew Sustain Energy Rev 77:515–524
27. Kandt A, Brown E, Dominick J, Jurotcih T (2007) Making the economic case for small-scale distributed wind—a screening for distributed generation wind opportunities. In: Wind power conference, Los Angeles, California, 3–6 June 2007
28. Etamaly AM, Mohamed MA, Alolah AI (2015) A smart technique for optimization and simulation of hybrid photovoltaic/wind/diesel/battery energy systems. In: 2015 IEEE international conference on smart energy grid engineering (SEGE). IEEE, pp 1–6
29. Ruhul Kabir M, Rooke B, Malinga Dassanayake GD, Fleck BA (2012) Comparative life cycle energy, emission, and economic analysis of 100 kW nameplate wind power generation. Renew Energy 37:133–141
30. Eltamaly AM, Mohamed YS, El-Sayed A-HM, Elghaffar ANA (2019) Analyzing of wind distributed generation configuration in active distribution network. In: 2019 8th international conference on modeling simulation and applied optimization (ICMSAO). IEEE, pp 1–5
31. Eltamaly AM, Farh HM (2015) Smart maximum power extraction for wind energy systems. In: 2015 IEEE international conference on smart energy grid engineering (SEGE). IEEE, pp 1–6
32. Eltamaly AM, Addoweesh KE, Bawa U, Mohamed MA (2014) Economic modeling of hybrid renewable energy system: a case study in Saudi Arabia. Arab J Sci Eng 39(5):3827–3839
33. Mohamed MA, Eltamaly AM, Farh HM, Alolah AI (2015) Energy management and renewable energy integration in smart grid system. In: 2015 IEEE international conference on smart energy grid engineering (SEGE). IEEE, pp 1–6
34. Wikipedia. World population. https://en.wikipedia.org/wiki/World_population. Accessed 28 Mar 2012
35. Bawah U, Addoweesh KE, Eltamaly AM (2013) Comparative study of economic viability of rural electrification using renewable energy resources versus diesel generator option in Saudi Arabia. J Renew Sustain Energy 5(4):042701
36. Karas KC (1992) Wind energy: what does it really cost? In: Proceedings of wind power '92, AWEA, pp 157–166
37. Eltamaly AM, Alolah AI, Farh HM, Arman H (2013) Maximum power extraction from utility-interfaced wind turbines. New Dev Renew Energy 159–192
38. EPRI (Electric Power Research Institute) (1989) Technical assessment guide. EPRI report: EPRI P-6587-L, EPRI, vol 1, Rev 6
39. Farh HM, Eltamaly AM (2013) Fuzzy logic control of wind energy conversion system. J Renew Sustain Energy 5(2):023125
40. Tande JO, Hunter R (1994) Estimation of cost of energy from wind energy conversion systems, 2nd edn
41. Eltamaly AM, Alolah AI, Abdel-Rahman MH (2011) Improved simulation strategy for DFIG in wind energy applications. Int Rev Model Simul 4(2)
42. Eltamaly AM (2007) Modeling of wind turbine driving permanent magnet generator with maximum power point tracking system. J King Saud Univ Eng Sci 19(2):223–236
43. Derrick A (1992) Development of the measure-correlate-predict strategy for site as assessment. In: Proceedings of BWEA
44. CEC (2003) Comparative Cost of California Central Station Electricity Generation Technologies. California Energy Commission 100-03-00IF. Prepared in support of the electricity and natural gas report 02-IEP-01, Aug 2003
45. El-Tamaly HH, El-Tamaly AM, El-Baset Mohammed AA (2003) Design and control strategy of utility interfaced PV/WTG hybrid system. In: The ninth international middle east power system conference, MEPCON, pp 16–18
46. Julian Bartholomy O (2002) A technical, economic, and environmental assessment of the production of renewable hydrogen from wind in California

47. Eltamaly AM (2000) Power quality consideration for interconnecting renewable energy power converter systems to electric utility. PhD diss, PhD thesis, Elminia University
48. El-Tamaly AM, Enjeti PN, El-Tamaly HH (2000) An improved approach to reduce harmonics in the utility interface of wind, photovoltaic and fuel cell power systems. In: Fifteenth annual IEEE applied power electronics conference and exposition (Cat. No. 00CH37058), APEC 2000, vol 2, pp 1059–1065. IEEE
49. Jangamshetti SH, Rau VG (1999) Site matching of wind turbine generators: a case study. IEEE Trans Energy Convers 14:1537–1543
50. Salameh ZM, Safari I (1992) Optimum windmill-site matching. IEEE Trans Energy Convers 7:669–675
51. Abdel-Hamid RH et al (2009) Optimization of wind farm power generation using new unit matching technique. In: 7th IEEE international conference on industrial informatics, Cardiff, UK
52. Lun I, Lam JA (2000) Study of Weibull parameters using long- term wind observations. Renew Energy 20:145–153
53. Aspliden CI, Elliot DL, Wendell LL (1986) Resource assessment methods, sitting and performance evaluation. World Scientific, New Jersey, pp 321–376
54. Diaf S et al (2008) Technical and economic assessment of hybrid photovoltaic/wind system with battery storage in Corsica island. Energy Policy 743–754
55. Bazyar R et al (2011) Optimal design and energy management of stand-alone wind/PV/diesel/battery using bacterial foraging algorithm. In: 8th international energy conference, pp 1–14
56. Lazou A, Papatsoris A (2000) The economics of photovoltaic stand-alone residential households: a case study for various European and Mediterranean locations. Sol Energy Mater Sol Cells 62:411–427
57. Kaabeche A, Belhamel M, Ibtiouen R (2010) Optimal sizing method for stand-alone hybrid PV/wind power generation system. In: Proceedings of SMEE, pp 205–213
58. https://www.wisegeek.com/what-is-replacement-cost.htm. Accessed 3 Mar 2012
59. Diaf S et al (2007) A methodology for optimal sizing of autonomous hybrid PV/wind system. Energy Policy 35:5708–5718
60. Nelson DD, Nehrir MH, Wang C (2006) Unit sizing and cost analysis of stand-alone hybrid wind/PV/fuel cell power generation systems. Renew Energy 31:1641–1656
61. Yang H et al (2008) Optimal sizing method for stand-alone hybrid solar–wind system with LPSP technology by using genetic algorithm. Sol Energy 82:354–367
62. Navaeefard A, Tafreshi S, Maram M (2010) Distributed energy resources capacity determination of a hybrid power system in electricity market. In: 25th international power system conference, PSC 2010, pp 1–9
63. Belfkira R et al (2009) Non linear optimization based design methodology of wind/PV hybrid stand alone system. In: Ecologic vehicles and renewable energies, Monaco, pp 1–7
64. Benyahia Z (1989) Economic viability of photovoltaic systems as an alternative to diesel power plants. In: Proceedings of the 9th European photovoltaic solar energy conference, pp 173–175
65. Skidmore EL (1986) Wind erosion climatic erosivity. Climatic change, vol 9. D. Reidel Publishing Company, pp 195–208

Printed by Printforce, the Netherlands